T0134527

Studies in Computational Intelligence

Volume 997

Series Editor

Janusz Kacprzyk, Polish Academy of Sciences, Warsaw, Poland

The series "Studies in Computational Intelligence" (SCI) publishes new developments and advances in the various areas of computational intelligence—quickly and with a high quality. The intent is to cover the theory, applications, and design methods of computational intelligence, as embedded in the fields of engineering, computer science, physics and life sciences, as well as the methodologies behind them. The series contains monographs, lecture notes and edited volumes in computational intelligence spanning the areas of neural networks, connectionist systems, genetic algorithms, evolutionary computation, artificial intelligence, cellular automata, self-organizing systems, soft computing, fuzzy systems, and hybrid intelligent systems. Of particular value to both the contributors and the readership are the short publication timeframe and the world-wide distribution, which enable both wide and rapid dissemination of research output.

Indexed by SCOPUS, DBLP, WTI Frankfurt eG, zbMATH, SCImago.

All books published in the series are submitted for consideration in Web of Science.

More information about this series at https://link.springer.com/bookseries/7092

Mohamed Abdel-Basset · Nour Moustafa ·
Hossam Hawash · Weiping Ding

Deep Learning Techniques
for IoT Security and Privacy

 Springer

Mohamed Abdel-Basset
Faculty of Computers and Informatics
Zagazig University
Zagazig, Egypt

Hossam Hawash
Faculty of Computers and Informatics
Zagazig University
Zagazig, Egypt

Nour Moustafa (ID)
School of Engineering and Information
Technology
University of New South Wales Canberra
Canberra, ACT, Australia

Weiping Ding
School of Information Science
and Technology
Nantong University
Nantong, China

ISSN 1860-949X   ISSN 1860-9503 (electronic)
Studies in Computational Intelligence
ISBN 978-3-030-89027-8   ISBN 978-3-030-89025-4 (eBook)
https://doi.org/10.1007/978-3-030-89025-4

This Springer imprint is published by the registered company Springer Nature Switzerland AG
The registered company address is: Gewerbestrasse 11, 6330 Cham, Switzerland

I dedicate this book to my readers for their curiosity to learn. My wife Ghofran, a divine presence and guidance constantly channels my creative energy to empower the world with my wisdom. My children (Aliaa and Laila) are inspiring me with their intuitive perspective, and their critique of the draft was instrumental in transforming the content for the audience.

—Mohamed Abdel-Basset

This book is dedicated to my parents, my wife who gives unconditional support for all the things I aspire to both vocationally and otherwise and to our children.

—Nour Moustafa

This book is dedicated to my parents, my wife who gives unconditional support for all the things I aspire to both vocationally and otherwise and to our children.

—Weiping Ding

To my adorable mom, E. Amera, and to my beloved sister, S. Hawash

—Hossam Hawash

Preface I

Internet of Things (IoT) is a ubiquitous technology for transforming everything we do and the infrastructure which supports us. From smart device to smart cities, companies, critical infrastructure, health care, smart grid, and wearables, which lead to vast amount of daily generated and processed data. This extraordinary scale, ubiquity, and interconnectivity also establish an environment where the confidence and integrity of these IoT applications turn out to be an overriding interest. One only has to look to the headings where attacks on vital infrastructure such as energy generation and distribution, liabilities in our automobiles, and malware in webcams, smartphones, and PCs which we bring into our homes underscore our joint vulnerability. Given the broad attack surfaces being formed and the irregularity between attackers wanting to uncover a particular weakness to take advantage of, whereas defenders should discover and close up all vulnerabilities, IoT generates an unrivaled set of security challenges. During journey of authors, we have had the pleasure of working with a lot of experts in their own fields. This well-timed book will build the readers knowledge about the IoT security challenges, the potential of deep learning to provide remedies for these challenges in the available IoT frameworks, and certain vertical applications. Please join us in the critical mission deep learning-based secure IoT applications.

Mohamed Abdel-Basset
Faculty of Computers and Informatics
Zagazig University
Zagazig, Egypt

Nour Moustafa
School of Engineering and Information
Technology
University of New South Wales at
ADFA
Canberra, ACT, Australia

Hossam Hawash
Faculty of Computers and Informatics
Zagazig University
Zagazig, Egypt

Weiping Ding
Ph.D, Full Professor, SMIEEE, Deputy
Dean of School of Information Science
and Technology
Nantong University
Nantong, China

Preface II

The topic of this book is deep learning for securing Internet of Things (IoT)—which is a subfield of artificial intelligence—concentrating on the general and challenging security problems in an IoT environment. Intelligence is driven only by learning the inherent representations from the observations acquired in IoT environment. This data-driven learning is incredibly general and may be applied to numerous everyday circumstances from playing games to autonomous driving of vehicles. Thanks to elasticity and generalization, the field of deep learning is evolving very rapidly and attracts lots of interest both from researchers attempting to enhance current or establish new techniques, as well as from specialists fascinated in resolving the cyber-physical security issues in the most effective mode. This book was written as an effort to fulfill the evident gap of workable and structured information about intelligent security solution in IoT environments. On one hand, there are countless research activities all around the world, new research papers are being published almost every day, and a large portion of conferences such as NIPS or ICLR is dedicated to deep learning methods for securing IoT applications. There are many huge research groups concentrating on deep learning solutions methods application in industry, robotics, health care, transportations, and others. The information about the recent research is broadly obtainable but is extremely specialized and abstract to be comprehensible without significant hard work. Even worse is the situation with the practical aspect of deep learning application, as it is not continuously apparent how to make a move from the abstract approach portrayed in the mathematical-heavy way in a research paper to a functioning application to real-world problem. This makes it challenging for somebody concerned in the area to obtain an insightful interpretation of approaches and concepts behind papers and conference talks. There are some very good blog posts about various deep learning aspects illustrated with working examples, but the incomplete presentation of a blog post tolerates the author to discuss only one or two approaches without designing a comprehensive planned picture and screening how diverse approaches are related to each other. This book is our effort to tackle this problem.

Zagazig, Egypt   Mohamed Abdel-Basset
Canberra, Australia   Nour Moustafa
Zagazig, Egypt   Hossam Hawash
Nantong, China   Weiping Ding

Acknowledgements

This book would not have been conceivable without the donations of several people. We would like to thank those who commented on our proposal for the book and helped plan its contents and organization: List some names. We would like to thank the people who offered feedback on the content of the book itself. Some offered feedback on many chapters: List some names. We also would like to thank our families as they never cease to support our work. We would like to thank the professors who instilled into us the value of education. I would like to express my gratitude to the people of Springer and this book's production team as they are inspiring and easy to work with. Finally, we would like to thank the institutions who always supported our teaching and research agenda, Zagazig University, Egypt.

About This Book

Who This Book is for?

The major aim audience are people who have some familiarity with Internet of Things (IoT) but interested to get a comprehensive interpretation of the role of deep learning in maintaining the security and privacy of IoT. A reader should be friendly with Python and the basics of machine learning and deep learning. Interpretation of statistics and probability theory will be a plus but is not certainly vital for identifying most of the book's material.

What This Book Covers

Chapter 1, "Introduction Conceptualization of Security, Forensics, and Privacy of Internet of Things: An Artificial Intelligence Perspective", discusses the main concepts behind the realization of reliable IoT including IoT security, IoT forensics, and IoT privacy. It also discusses the potential of artificial intelligence, machine learning, and deep learning for securing an IoT system. The derivation of deep learning and the taxonomy of classifying deep learning models are also discussed in this chapter.

Chapter 2, "Internet of Things, Preliminaries and Foundations", discusses the definition of IoT system and the layered architecture of IoT system and provides a fine-grained definition for layers of IoT system, along with their responsibilities, elements, and rankings. The chapter also argues important concepts and computing paradigms greatly related to the IoT system, such as cloud computing, fog computing, and edge computing.

Chapter 3, "Internet of Things Security Requirements, Threats, Attacks, and Countermeasures", provides a thorough discussion about different attack surfaces and the potential vulnerabilities in IoT systems. This entails different threats and attacks across different tiers or layers of IoT systems. Then, a taxonomy for dividing the attack surfaces into multiple categories is presented along with the primary security threats and possible vulnerabilities per each category. Furthermore, a novel attack surface is triggered by the mutually dependent, interrelated, and co-operating environments of IoT systems.

Chapter 4, "Digital Forensics in Internet of Things", presents a compact discussion of the theoretical approaches, essential challenges, and research developments of digital forensics in IoT-based system. Besides, it figures out the requirement for standardization the forensics procedures, as this is considered as a crucial phase toward high-level quality cross-jurisdictional forensics articles and ideal procedures for assuring cyber-physical security.

Chapter 5, "Supervised Deep Learning for Secure Internet of Things", presents an in-depth discussion for the supervised deep learning approaches for maintaining the security of IoT system. This includes the state-of-the-art supervised algorithms for securing IoT systems, the main characteristics, corresponding advantages and disadvantages, and the corresponding role in IoT security and also provides detailed tabular comparisons and figures that indicate helping the readers to grasp the learned lessons.

Chapter 6, "Unsupervised Deep Learning for Secure Internet of Things", presents a comprehensive discussion for the learning from unlabeled IoT data using unsupervised deep learning approaches for the purpose of securing the IoT environments. This involves the state-of-the-art unsupervised algorithms along with their main characteristics, the corresponding advantages and disadvantages, and some use cases in IoT security and also provides a detailed tabular comparison and figures that help the readers obtain full understanding of the methodology of unsupervised learning.

Chapter 7, "Semi-supervised Deep Learning for Secure Internet of Things", presents a detailed discussion about learning from partially labeled data using the semi-supervised deep learning approaches for maintaining the security of IoT system. This includes the state-of-the-art supervised algorithms for securing IoT systems, the main characteristics, corresponding advantages and disadvantages, and the corresponding role in IoT security. Furthermore, it provides a detailed tabular comparison and figures that aim to help the readers grasp the learned lessons.

Chapter 8, "Deep Reinforcement Learning for Secure Internet of Things", introduces and discusses the potential of deep reinforcement learning for maintaining the security of IoT system. This includes the reinforcement learning supervised algorithms, the key similarities, and differences, the related advantages, and disadvantages, and the condition of usages securing IoT environment. Moreover, illustrative diagrams are also presented to emulate the workflow of different reinforcement learning frameworks to assist the readers to completely comprehend the role of reinforcement learning in securing the IoT environments.

Chapter 9, "Federated Learning for Privacy-Preserving Internet of Things", discusses privacy considerations of applying intelligent solutions in IoT-based system. The federated learning paradigm is extensively discussed, categorized, and analyzed for privacy preservation. Moreover, it also deliberates the role of differential privacy, multi-parity computing, and other methods of privacy preservations.

Chapter 10, "Challenges, Opportunities, and Future Prospects", introduces the current and most recent challenges facing the deep learning community when designing privacy and security preserving IoT system and also presents possible solutions that could be investigated by the academic or industrial community in near future. Furthermore, the convergence of deep learning approaches and IoT technologies is still in its early phases paving the way toward more reliable and trustworthy IoT solutions which are figured out as a part of future work in either academia or marketspace.

Contents

Chapter 1
Introduction Conceptualization of Security, Forensics, and Privacy of Internet of Things: An Artificial Intelligence Perspective

The Internet of Things (IoT) is a rapidly evolving technology that empowers billions of globally distributed physical things to be interconnected over the internet to capture, collect, exchange, and share a wide variety of vast amounts of data. These physical things incorporate all of the connectable devices ranging from conventional household devices to complex industrial devices [1]. The emergence of INDUSTRY 4.0 has revolutionized the industrial domain globally with a large and ever-increasing number of embedded devices, and computerized chips because of the expanding availability, affordability, microprocessors, sensors' capacity, and omnipresent communication technologies. Connecting such a very large number of heterogeneous objects and implanting sensors to them leads to some degree of digital intelligence at devices that, empowering them to communicate instantaneous data with no human intervention. The IoT empowering the infrastructure of the globe to be more intelligent and reactive, combining the physical and digital domains. Therefore, the IoT has been receiving great interest and wide adoption in almost all fields: education, manufacturing, healthcare, business, entertainment, energy distribution, domestic, smart-cities, transportations, tourism, and even research [2].

Regrettably, industry, academia, and people are struggling to integrate the stream of rapid commercialization with rare consideration to the security and the privacy of IoT data and devices. For example, think about an IoT-enabled world that combines things like wearables, smartphones, smartwatches, smart TVs, smart refrigerators, smart vehicles, implantable health devices, industrial robotics systems, and simply everything capable to be connected over networks. Generally, most of these industries have never posed any interest in security, forensics, or privacy concerns over the last decades [3]. Given the intense competition to be more feasible along with profitable products and services, nonetheless, the current industries discover themselves in the critical phase as they do not realize the secure and privacy-preserving method

Electronic Supplementary Material The online version of this chapter (https://doi.org/10.1007/978-3-030-89025-4_1) contains supplementary material, which is available to authorized users.

M. Abdel-Basset et al., *Deep Learning Techniques for IoT Security and Privacy*, Studies in Computational Intelligence 997, https://doi.org/10.1007/978-3-030-89025-4_1

to develop, deploy, and operate. Such negligence can conceivably threaten the IoT users and in turn interrupt the vibrant ecosystem. Hence a lot of industries, clients and business owners, service providers, and infrastructure operatives are rapidly realizing themselves facing a variety of intrusions and security violations. The effort made to turn standard devices into smart devices is establishing a frenzy of chances for cyber-criminals, intruders, data poisoners, and malicious agents [4].

The threat disclosed by these IoT things not just influences the security of IoT systems, but also leads to the catastrophic corruptions eco-system involving services, applications, data centers, websites, individual residents, and social networks, through self-controlled smart gadgets as robot networks (botnet). In other words, negotiating a specific element and/or communication channels in IoT systems is enough to disrupt a large portion of the entire IoT network. Beyond the vulnerabilities of IoT systems, the attack vectors are considerably developing in terms of variety, and complication [5].

In this wave of interest, the security and privacy of IoT-based system embody one of the major weak points holding back the approval of the IoT in different fields, as the IoT devices are habitually poorly secured [6] and hence easy targets for malware to exploit them to trigger destructive or interference attacks. It is therefore obvious that absence of security procedures, the IoT-based system remains unable to provide its complete capabilities. As A Result, extra consideration has to be given to the investigation of those attacks, their discovery, contamination prevention, countermeasures, and system recovery following the attacks.

In the nutshell, there will forever be some people striving to break into and negotiate computer systems and smart devices. Worse, with the proliferation of IoT, the incentive might expand to inflicting physical injury or killing somebodies. Presently, a simple keystroke led to critical consequences such as (1) it can protect human life when appropriately designing a pacemaker, (2) it can enable exploring the outer space, (3) it can destroy economies of large corporations or even countries, (4) it can deactivate a vehicle's braking system, (5) it can stumble a nuclear research system, etc. [7]. Therefore, intelligent solutions for preserving the security and privacy of IoT-based system are of vital importance, however, before probing into actual facets of maintaining the security and privacy of IoT, the residual part of this chapter will discuss the following:

- *Cyber-Physical Internet of Things*
- *Security of Internet of Things*
- *Forensics of Internet of Things*
- *Privacy of Internet of Things*
- *Artificial Intelligence for Secure Internet of Things*
- *Summary and Learnt Lessons.*

1.1 Cyber-Physical Internet of Things

The Internet of Things (IoT) is a dynamic universal network infrastructure with self-organizing facilities founded on interoperable and standard internet protocols, enabling flexible communication between physical things to share information sharing and achieve cooperative decision-making [8]. In IoT, every connected thing has a standalone identity, and all things are interconnected over a diversity of communication networks to promote dynamic information collaboration between them irrespective of whether they are industry or national oriented.

Cyber-physical system (CPS) was firstly presented in 2006 by Helen Gill [9], give an emphasis to the in-depth amalgamation of a variety of information technologies, such as sensing technology, hardware technology, software technology, implanted technology, and micro-processing technology with the main aim to realize the extremely self-directed and synergistic information-ization competences, instantaneous and elastic feedback, and constructive cycle between the information and physical worlds [10]. The boundary of CPS is very large fusing an extensive range of engineering specialties, practices, applications, and pertinent affairs (i.e., digital design, thermodynamics, engineering dynamics, etc.) required by their corresponding practitioners [6].

How the CPS differ from the IoT system? According to the IEEE, the fundamental difference is that a CPS encompassing connected physical devices, monitoring process, and control systems that are optionally connected to the Internet. In other words, the CPS can be inaccessible from the Internet and remain operative and achieving the corresponding business goals. From an IoT perspective, the things in the IoT system inevitably and by definition, are connected to the Internet, and use some applications, services, system functionality to accomplish targeted business goals. Hence, the primary concentrate the interactions between objects over the Internet where openness, globalization interoperability, and socialization greatly support the concept of IoT [8]. Since all things of IoT can be considered as a subset of CPS with only one condition which is internet connectivity, then, the IoT is considered as a subset of standard CPS, as shown in Fig. 1.1. In addition, another subset of CPS is the industrial internet, which exposes the prospective future developments of industries

Fig. 1.1 The relationship between the cyber-physical system, internet of things (IoT), industrial internet, and industrial IoT (IIoT)

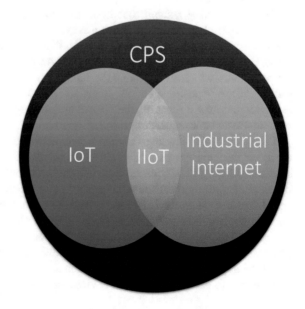

grounded on emergent technologies (i.e., 5G, 6G). The intersection between IoT and industrial internet brings us to a new form of IoT called industrial IoT (IIoT). Furthermore, when all elements of CPS are connected to the internet, then it could be referring to it as a cyber-physical IoT or might also be called an IoT system [11].

1.2   Security of Internet of Things

Recently, there is a large misunderstanding between IoT security and conventional cybersecurity, however, they completely differ from each other either conceptual or practical perspectives. On one hand, cybersecurity is generally identified as a technology pile of methods, protocols, and procedures to protect computing systems (software, application, services), network infrastructure, and data against malevolent attacks, intrusions, unauthorized access, and other categories of deliberate and/or accidental damage. According to this, the prime emphasis of cybersecurity is to safeguard the data and information, meanwhile ignoring the security requirements of the physical hardware device or as well as the interactions physical world with different elements IoT system [2]. On the other hand, the security of IoT is a more broad, complex, and challenging task due to larger attack surfaces introduced by the IoT system, and the increased vulnerability of physical things at these surfaces. Hence, IoT security can make use of cybersecurity combined with other security procedures to be jointly followed to guarantee thorough consideration of all security facets. Hence, IoT security can be defined as the fusion of cybersecurity, physical device security, and security of technologies, control processes, and procedures essential to protect different elements of IoT-based systems from a broad spectrum of IoT security attacks [5].

Broadly speaking, a significant aspect that differentiates the concept of IoT security from cybersecurity is what is known as cyber-physical systems through the industrial and academic community. For example, in a power station where a temperature sensor is under remote attack, then it can erroneously generate very high-temperature measurements, which might trigger the control system to shut down the whole power station. In the opposite situation, where the sensor mistakenly generates a lower measurement than it must be, the control system might take more serious actions, with unpleasant consequences [12].

Digital control of physical operations over networks makes the IoT distinct in the sense that the underlying security solution is not restricted to elementary information assurance ideologies of non-negation, integrity, confidentiality, etc., however that of physical resources and machines that generate and accept that IoT data in the physical world. In other words, the IoT-based system has extremely genuine analog and physical components. The IoT devices are considered physical objects, most of which are having some security concerns. This, compromising these devices would cause physical damage to properties and injuring individuals. The topic of IoT security, so, is not the application of a specific, constant set of standard security procedures as they operate to network-connected devices and servers. It necessitates a distinctive and intelligent solution for each application and service to which the IoT devices contribute.

1.3 Forensics of Internet of Things

The discipline of Digital Forensics (DF) is considered as a part of the conventional forensics science that mainly considers the discovery and analysis of digital or computerized data. The DF specialists deal with the recognition, compilation, retrieval, analysis, and maintenance of digital evidence discovered on a variety of electronic devices [13].

With the emergence of IoT technology and the strong connection between human individuals and IoT, civilian and criminal investigations or internal probes should take IoT into consideration. The security breaches of IoT systems could be exploited to remotely take the control of the systems, for instance, maliciously control the vehicle's braking system to make an accident. Thus, the IoT system witnessing an urgent necessity for IoT forensics research to support in deciding the who, what, where, when, and how for cases. From the digital forensic standpoint, the IoT environment comprises a plentiful set of objects that may possibly benefit investigations, while the digital forensic investigation in the IoT paradigm has been developed to accommodate characteristics of IoT, leading to a new concept known as IoT Forensics [14].

The main difference between IoT Forensics and digital forensics can be observed in terms of evidence sources [12]. Different from standard digital forensics, in which the laptops, computers, smartphones, smartwatches, cloud servers are employed as a normal substance of inspection, the sources of evidence within IoT Forensics can be much extra wide-spreading containing medical implantations in creatures, infant nursing systems, traffic lights, and In-Vehicle Infotainment systems [15]. Preserving the security of overall IoT devices, communications, and storage within different layers of the IoT network is essentially difficult if not impossible to realize. If an incident happens, one of the first tasks that forensics professionals execute is to define the scope of the compromise. Nevertheless, different from standard IoT security procedures, the target of IoT forensics is not reducing the amount of incurred destruction, yet to distinguish the origin of attacks or the responsibilities of the various factions [16]. Though existing forensic practices and instruments still demonstrated benefits at some phases of forensics in IoT territory, however, serious requirements are still imposed to update current implements, processes, and legislation to deal with distinctive and evolving traits of IoT.

1.4   Artificial Intelligence for Secure Internet of Things

Artificial intelligence (AI) was firstly presented in 1950 as an answer to the question "could the computers think?" Then, Artificial intelligence (AI) was defined as a discipline of computer science that mainly emphasis tackling and solving problems that are rationally challenging for human individuals but comparatively simple for computers. This refers to problems that can be designated according to a group of official, arithmetic, scientific rules. Intrinsically, AI is an extremely broad discipline that involves learning-based methods and many categories of methods that do not entail any learning process [17]. Indeed, for a relatively long-time computer science professionals thought that human-level AI could be accomplished merely by designing a satisfactorily substantial set of unambiguous rules for processing and handling the knowledge. This concept is later identified as "symbolic-AI", which was the leading concept within the artificial intelligence community from 1950 till 1980. The symbolic AI achieved its peak celebrity throughout the prosperity of the expert systems by 1980. Though symbolic-AI showed appropriate to provide a solution for rational and formally defined challenges (i.e., playing chess), but, it fails to be awkward to discover explicit rules to solve more complicated, unstructured, and ambiguous problems.

The actual challenge for artificial intelligence demonstrated to be finding a solution for the problem that can be easily performed by human individual but is difficult to be formally described by human individuals, i.e., the problems that are intuitively solved, and automatic in the sense, such as faces recognition, object detections, word recognition, etc. [18]. In this regard, the security vulnerabilities of IoT-based system can be considered and formulated as a problem that necessitates an intelligent solution to protect the devices and information from potential IoT threats [2].

Artificial intelligence provides a variety of techniques for exploring and learning the legitimate and malicious patterns from IoT data streams generated as a result of the interaction between different IoT components and physical devices. The generated data from different sectors of the IoT system can be aggregated and investigated by a single or multiple AI approached to discriminate normal patterns of interaction from abnormal ones, thereby recognizing IoT threats or malicious activities at early phases, before causing any damage to any element of the IoT system. Moreover, Artificial intelligence could be crucial for forecasting new IoT threats, which almost are a variation of preceding threats, for the reason that they can smartly envisage future anonymous threats based on the experience gained from current samples. Furthermore, artificial intelligence might help the IoT system to intelligently select the most appropriate defense mechanism or countermeasures based on the acquired knowledge about different threats. Thus, IoT-based systems should have a switch from just enabling secure communication between different IoT parities to security-based intelligence for reliable and trustworthy IoT systems [19]. Figure 1.2 illustrates the potential of artificial intelligence approaches to maintain the security and privacy of IoT systems. Among the existing artificial intelligence paradigms,

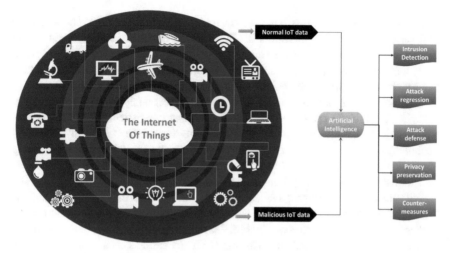

Fig. 1.2 Visual representation of the potential role of artificial intelligence in securing the internet of things

machine learning is the most broadly used paradigm, which has great achievements in different application domains, even in security-related applications.

1.4.1   Machine Learning

The concept of "machine learning (ML)" is a large subfield of artificial intelligence, that was first introduced by Arthur Samuel [20], which defined as a computer system (which is playing checkers game) having the ability to learn from large amounts of historical data, using a set of self-adapting computational algorithms. In fact, the hard programming of computer algorithms makes it unable to cope with the dynamically changing requirements and continuously updated system conditions. On the contrary, the machine learning algorithms are trained instead of purely programmed, the ideology of learning from previous experiences enables the machine learning algorithms to help in the decision-making process and forecasting tasks, which can be accomplished by building a self-optimizing algorithm from input data.

For extra clarification for the notion of "learning", Mitchell [21] provided the extensively quoted explanation: *"A computer program is said to learn from experience E with respect to some class of tasks T and performance measure P, if its performance at tasks in T, as measured by P, improves with experience E."*

The prosperity of machine learning began in 1990 [22]. Prior to this time, knowledge-based and rule-based paradigms (i.e., expert systems, inductive logic programming, and others) were dominating the view of artificial intelligence depending on complex human-comprehensible symbolic representations of reasoning. With the emergence of stochastic estimate and statistics philosophy, machine learning paradigms have reattracted the attention of researchers leading to a variety of valuable probabilistic algorithms. Both academia and industry go aboard on developing data-driven approaches for scrutinizing and analyzing a relatively larger amount of data and strained to learn and extract insights that might help to accelerate or improve the decision-making in different application domains.

Though the tight relation between ML and statistical analysis, but, it varies from statistics in multiple critical aspects. Different from statistics, ML tends to deal with relatively larger, more complicated datasets for which "classical" statistical methods (i.e., Frequentist statistics, Bayesian statistics) would basically be extremely unpractical to be viable. Also, ML is a practical field of AI in which the concepts can be established in experimental manners rather than only theoretical proofs [7, 23].

With the design of progressive learning approaches [24], the interest in machine learning has transferred from "the learning as an objective" to "the learning as functionality". Unambiguously, machine learning algorithms have been no longer sightlessly operate to emulate the learning ability of humans, as an alternative, they emphasize more on intelligent and task-specific data-driven analysis. At the present time, with plenty of data and the regular interaction between data exploitation and data exploration operations, machine learning algorithms have been achieving great success in various topics i.e., data mining, information retrieval, smart control, etc. IoT networks seek to offer pervasive applications or services for a large number of individuals centrally or in a distributed manner. However, security and privacy are the main concerns to judge the reliability and trustworthiness of these applications or even the entire IoT system. As a remedy, machine learning helped to improve IoT security against different categories of threats and is considered as a cornerstone of securing the IoT-based systems taking into account their distributed and dynamic nature.

In this regard, support vector machine (SVM), Decision Tree (DT), and Random Forest are extensively employed to discriminate normal IoT samples from Attacked IoT samples, and they demonstrated high classification accuracy [7]. In another way, SVM and k-means algorisms are employed to detect mischievous IoT nodes by learning the node behaviors from accumulated data about all nodes [25]. Additionally, instead of directly applying the ML algorithms directly to the IoT data, other studies concentrate on improving the performance of ML algorithms by designing a robust feature engineering technique. For example, a wrapper-based feature selection method (called CorrAUC) was introduced in Ref. [26], to precisely select and filter up the significant IoT features to be passed to the ML algorithms rather than the original set of features, which demonstrated as effective for improving the efficiency of different ML algorithms in detecting BoT-Net traffic in IoT data.

Despite the great success of conventional machine learning algorithms, however, still suffering from critical shortcomings that limit their applicability in IoT environments.

- The efficiency of the machine learning algorithms is heavily dependent on the selection of a proper group of features; hence, feature engineering techniques become an essential part having a vital role in the design of any machine learning solution. Unfortunately, the tasks of feature engineering (i.e., feature extraction, feature selection) are complex and difficult to accomplish, also they necessitate much computational overhead in terms of memory usage or processing capabilities.

- Generally, Machine learning algorithms necessitate sufficient time to be trained and possibly to efficiently solve the underlying task. Typically, this necessitates additional power processing, and memory resources, which is not always available in IoT devices.
- Machine learning algorithms are susceptible to the aphorism "garbage in, garbage out." This aphorism implies that the quality of input data decides the superiority of the output, thereby the performance of the algorithm. Thus, when the IoT data are noisy or weakly annotated, then the performance of the ML algorithm would be affected by these flaws.
- With the emergence Big Data industry tightly connected to IoT, high dimensionality data becomes a chief problem in that limits the performance of standard ML algorithms even they perform professionally on conventional low dimensional data. This phenomenon is broadly known as the "Curse of Dimensionality". Despite the efforts made to address this problem, but it still a major limitation for ML algorithms.
- Another key limitation of machine learning algorithms is the lack of interpretability as it is difficult to clarify the way these algorithms reach their final decisions or conclusions. Hence, the standard Machine learning algorithm is viewed as a black box receiving some inputs and generating a particular output without explaining the factors or the reasons that lead to this output. For example, when an ML algorithm classifies an IoT traffic sample as an attack, the algorithm cannot provide an explanation for the motives behind this classification decision.

An empirical case study for applying ML for IoT security is given in the supplementary materials of this chapter.

1.4.2 Deep Learning

In a stream of efforts to address the previously mentioned limitations of machine learning techniques especially in IoT environments, Deep learning emerged as a subcategory of machine learning motivated by the neural network algorithms in the working methodology of the human brain. Deep learning has been becoming so popular because of the great computational improvements and the increasing amount of data streams generated through IoT networks. With the big data volume, the deep learning models overcome the limitation of standard machine learning algorithms, and it has already been revolutionizing and expansively applied in numerous interdisciplinary scientific domains including computer vision, medical image analysis,

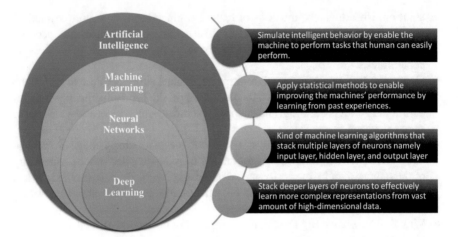

Fig. 1.3 Visual representation of relationship between artificial intelligence (AI), machine learning (ML), neural network (NN), and deep learning (DL)

natural language processing (NLP), human–computer interactions, and other applications. Figure 1.3 illustrates the relationship between deep learning, neural networks, machine learning, and artificial intelligence.

1.4.2.1 Biological Neurons

Prior to going forward, the concept of the neuron is firstly explored along with its working methodology within the human brain, which is considered as the main motive behind the artificial neurons, thereby the networks. A biological neuron could be described as the fundamental computational element of the human nervous system and brain. The human brain comprises roughly a hundred billion neurons, each of which is linked to each other through a biological construction known as a synapse. This synapse is responsible for accepting electrochemical input from the outer environment, sensory organs for transmitting dynamic directives to the physical muscles, and for carrying out other activities. In addition to the connecting synapses, the biological neuron is principally constituted of three main elements. A biological neuron could also accept inputs from the surrounding neurons through a branch-shaped formation known as a dendrite. The neuron's inputs could be boosted or weakened by weighting them based on their significance, then, they are collectively added or combined in some way at the body of the cell known as soma. The combined inputs at the cell body are subsequently processed and transferred to the other neurons across the axons. To sum up, the mission of accepting the forthcoming electrochemical information is accomplished via dendrites, then the manipulation and processing typically performed at the Soma. Forthcoming signals could be either excitatory—which indicates that it motivates the neuron to fire (produce an electrical pulse)—or

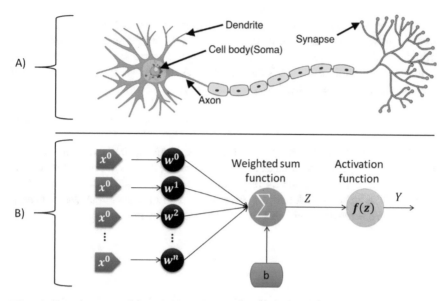

Fig. 1.4 Illustration of biological neuron (**a**) in comparison with artificial neuron (**b**)

inhibitory—which implies it encourage the neuron to keep stable without firing. In other words, the biological neuron fires if and only if the total electrochemical signal obtained at the Soma surpasses a specific level or threshold value. A single biological neuron might have multiple sets of dendrites and might possibly accept many thousands of input signals. the neuron's decision to be enthused into firing an impulse hinge on the totality of overall the received inhibitory and excitatory signals as well as firing threshold value. When the neuron fires, the nerve impulse transmitted over the axon which divides into several branches and builds bulbous swellings recognized as axon terminals, where the connections to the dendrites are made through a synapse (see Fig. 1.4a).

1.4.2.2   Artificial Neurons

The artificial neuron represents a mathematical model of a biological neuron, broadly identified as perceptron [27]. Like the dendrites that accept electrochemical signals from the surrounding biological neurons, the perceptron represents these input signals using a set of numerical values. The modulations of electrochemical signals at the synapses are also modeled in the perceptron by multiplying each numerical input by a weight value to represent the significance of this input to the output of the perceptron. Inspired by the firing phenomenon of biological neuron i.e., the total input intensity exceeds the predefined threshold, the perceptron computes the weighted sum of the numerical inputs to characterize the total intensity of inputs, then apply an activation function on the result to decide its output. As with biological neurons, the output of the perceptron can be passed as an input to other artificial neurons. Figure 4b display an illustrative diagram of the workflow of a single perceptron. Given the artificial neurons with three inputs x_1, x_2, x_3, and the corresponding weights dented as w_1, w_2, w_3, while its output is denoted as y. At first stage, the weighted sum of the inputs is computed as follows:

$$z = x_1 \cdot w_1 + x_2 \cdot w_2 + x_3 \cdot w_3 \qquad (1.1)$$

When the input x_1 is more important than the input x_1, the weight w_1 ought to be larger than w_2. For convenience and generalization, the set of input values can be represented as a vector x, and corresponding weights can be denoted as vector w. Then, the bias value is added to the weighted sum to guarantee that even when all the inputs are zeros there will still be activation in the neuron.

$$z = \sum_{i=1}^{n} x_i \cdot w_i + b \qquad (1.2)$$

where n represents the number of inputs, and b is a constant representing the systematic prediction error that helps the model to have the freedom to best fit the data thereby perform better. Finally, the **activation** function $f(z)$ is applied to define the way the weighted sum z of the input going to be transformed into perceptron's output. The activation function is sometimes called the *transfer function*. Thus, perceptron output is defined as follow:

$$Y = f(z) \qquad (1.3)$$

Various activation functions are nonlinear and hence might be termed as the "*nonlinearity function*". A perceptron without an activation function is fundamental to acts as a linear regression model. Thus, the main objective of activation/transfer functions is to promote the non-linear transformation to the input enabling the perceptron to learn from deep representations and thereby perform more intricate tasks. Table 1.1 present the main types of activation functions, their definitions, and mathematical formulas.

Figure 1.5 shows the diagram of the most popular activation functions.

1.4.2.3   Artificial Neural Networks

Since the construction of biological neurons are simple, it is not possible to just make use of a single biological neuron to execute complicated tasks. Hence, the human brain contains billions of connected biological neurons, which are organized into layers, constituting a large network.

Motivated by this, the perceptrons are staked into successive layers to construct an artificial neural network (ANN), where all layers are connected to enable the

Table 1.1 List of common activation functions for artificial neuron

Activation functions	Study	Description	Formula
Sigmoid	[28]	– Known as a logistic function – The sigmoid function is a broadly applied activation function that scale the input to a range between 0 and 1 – It is monotonic since it is either completely non-increasing or non-decreasing – It is differentiable since the slope of the curve is attainable at any two points	$f(z) = \frac{1}{1+e^{-z}}$
Hyperbolic tangent (Tanh)	[29]	– The function accepts real-valued input and outputs values in the range of [-1, 1] – It is monotonic – It is differentiable	$f(z) = \frac{1-e^{-2z}}{1+e^{-2z}}$
Rectified linear unit (ReLU)	[30]	– It maps the input into an output a value from o to infinity – It is essentially a **piecewise** function	$f(z) = \begin{cases} 0, & for\, z < 0 \\ z, & for\, z \geq 0 \end{cases}$

(continued)

Table 1.1 (continued)

Activation functions	Study	Description	Formula
Leaky ReLU	[31, 32]	– A variant of the ReLU function that resolves the issue of dying ReLU – The negative input has a small linear slope instead of transforming every negative input to zero – The value of variable σ can be set as a parameter of the model to learn its optimal value (called as **Parametric ReLU**) Instead of setting some random values (called as **Randomized ReLU**)	$f(z) = \begin{cases} \propto z, \ for\, z < 0 \\ z \quad for\, z \geq 0 \end{cases}$
Exponential linear unit (ELU)	[33]	– A Variant of ReLU that provides a small curvy slope for negative inputs. But – It is not continuously differentiable regarding its input when the value $\propto$ is not 1	$f(z) = \begin{cases} \propto (e^z - 1), \ for\, z < 0 \\ z, \quad\quad for\, z \geq 0 \end{cases}$

(continued)

Table 1.1 (continued)

Activation functions	Study	Description	Formula
Swish	[34]	- It is a non-monotonic function - σ represent the sigmoid function - It can be reparametrized with β that offer a good interpolation between linear and nonlinear function	$f(z) = z \cdot \sigma(z)$ $f(z) = 2z \cdot \sigma(\beta z)$
SoftMax	[35]	- It is a generalization of the sigmoid function - In a multi-class problem, it calculates the probabilities for each output class z_j, where the total probability of all classes equal to 1	$f(z_j) = \dfrac{e^{z_j}}{\sum_j e^{z_j}}$
Softplus	[36]	- It smooth approximation to the ReLU function to constrain the output of a to be positive - It can be reparametrized with β such that it returns to the linear function when the $z \times \beta$ exceeds the predefined solution	$f(z) = \log(e^z + 1)$ $f(z) = \dfrac{1}{\beta}\log(e^{\beta \cdot z} + 1)$
Softsign	[37]	- It constrains the output to range from -1 to 1	$f(z) = \dfrac{z}{\mathrm{abs}(z)+1}$

(continued)

Table 1.1 (continued)

Activation functions	Study	Description	Formula
Scaled ELU (SELU)	[38]	– self-normalizing activation to promote high-level abstract representations – It enables the model to converge towards zero mean and unit variance – where α and s are pre-specified constants ($\alpha = 1.67326324$ and $s = 1.05070098$)	$f(z) = s \cdot \begin{cases} z, & for\ z < 0 \\ \alpha \cdot (e^{\beta \cdot z} - 1), & for\ z > 0 \end{cases}$
Exponential	/	– It just computes the exponential of the input	$f(z) = e^z$
Gaussian error linear unit (GELU)	[39]	– Broadly used n NLP, specifically Transformer models – Escapes vanishing gradients dilemma	$f(z) = z * \Phi(z) = z * P(Z \le z)$ $\approx 0.5\left(1 + \tanh\left[\sqrt{\frac{2}{\pi}}(z + 0.044715z^3)\right]\right)$
Continuously differentiable ELU (CELU)	[40]	– It is a parameterization of ELU to make its derivative bounded regarding the input – It encompasses linear and ReLU activation as a special case – It is scale regarding $\propto$	$f(z) = \begin{cases} \propto (e^{z/\propto} - 1), & for\ z < 0 \\ z, & for\ z \ge 0 \end{cases}$

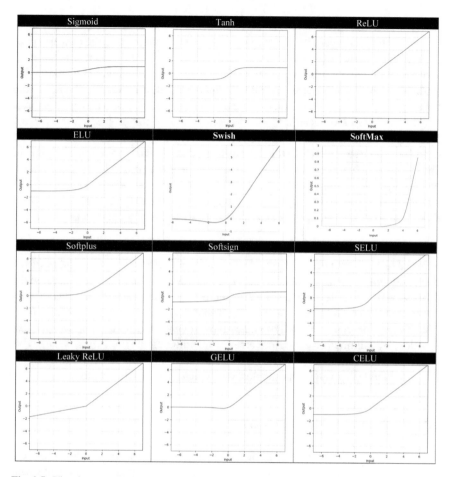

Fig. 1.5 Visual representation of behavior of popular activation function where the input value represented at X-axis, and the output values represented at Y axis

information to pass from one layer to another. Generally, the architecture of ANN consists of three primary layers, namely the Input layer, Hidden layer, and Output layer. Each layer consists of a set of neurons, and the neurons per layer have interaction with all the neurons in either the previous or next layer. Nonetheless, there is no interaction between the neurons in the same layer as they have no connections to link them, while the neurons in the neighboring layers have intermediate edges connecting them.

As the name implies, the Input layer takes the responsibility of feeding the incoming inputs to the network, where the number of input neurons is equal to the number of the network's input values. Each input value contributes to a different degree to the target prediction. Nevertheless, there is no computation to be accomplished at this layer, as it can be considered as a gate to pass information from the outer globe to the network. On the other hand, the hidden layer represents any layer between the input layer and the output layer, its main function is the processing of the incoming inputs from the input layer. Specifically, the hidden layer is accountable for modeling complicated relationships between input and output by extracting the hidden patterns from the data and learning the valuable representation necessary for output calculation. The choice of the number of hidden layers has no standards, but it was determined according to the complexity of the problem to be solved. Stacking a large number of hidden layers in ANN brings us to a network termed a deep neural network (DNN), which is considered the foundation of the discipline of deep learning. Finally, the Output layer receives the processing information from the hidden layer and computes the output of the network accordingly. The number of outputs perceptrons in this layer is determined according to the underlying problem that the network seeks to solve, or simply according to the target that seeks to estimate [41].

The learning of ANN means updating the weights of every perceptron in such a way that network correctly predicts the targeted output. The correctness of the predicted output usually measured using predefined cost functions that measure the difference between the actual and estimated output. The update of weights of each perceptron is accomplished using an algorithm known as a gradient descent [42]. The gradient descent iteratively updates the network weights to optimize the cost function, by minimizing the output error or maximizing the prediction accuracy. A detailed description of the learning implementation, training, testing of the simple ANN and DNN is given in the supplementary materials of this chapter.

1.4.2.4 Deep Learning: Definitions

According to Raschka and Mirjalili [23], deep learning can be defined as "a subfield of machine learning that is concerned with training the artificial neural network with many layers efficiently". Similarly, Chollet [43] defined deep learning as "a specific subfield of machine learning, a new take on learning representations from data which puts an emphasis on learning successive "layers" of increasingly meaningful representations". Besides, Goodfellow et al. [35] Considered deep learning as a learning technique that tolerates the computing environment to evolve based on the data and experience and defined it as "a particular kind of machine learning that achieves great power and flexibility by learning to represent the world as a nested hierarchy of concepts and representations, with each concept defined in relation to simpler concepts, and more abstract representations computed in terms of less abstract ones".

The introduction of deep learning alleviates the need to handcraft the input features by applying multi-layer ANN to learn the complicated inherent features and multi-level abstract representation. Hence, different from the prevailing machine learning algorithms, the deep learning networks do not necessitate a manual feature engineering technique as the network layer take the responsibility extracting and learning the valuable combination of features necessary to accomplish the undergoing task efficiently.

The applications of deep learning span different fields with different modalities of data. For example, computer vision applies a variety of deep learning algorithms to process image and/or video data for different purposes. Besides, medical diagnosis applications employ deep learning models process to process clinical measurements and radiographic images of patients to provide doctors with insightful diagnostic decisions about the patients. The NLP applications make use of deep learning algorithms to manipulate textual data for different purposes such as text summarization, translation, sentiment analysis, etc. Large enterprises such as Google, Microsoft, Nvidia, Netflix, Amazon, and Tesla have made a great shift toward the development of intelligent applications or services based on the deep learning models. According to this, deep learning can be viewed as a rapidly evolving field to deliver intelligence to a wide variety of domains, while alleviating the restrictions obscuring the standard machine learning approaches. In this regard, deep learning can be viewed as a promising solution to satisfy the evolving security and privacy requirements of IoT-based systems [44]. This solution entails attack detection, attacks forecasting, defense mechanism, and privacy preservations, etc. Therefore, this book seeks to investigate the potential of deep learning models as the state-of-the-art algorithms for designing interesting security, forensics, and privacy-preservation solutions for improving the reliability and trustworthiness of IoT-based systems.

1.5 Taxonomy of Deep Learning Approaches

The current deep learning approaches are typically classified into four main categories including supervised deep learning, unsupervised deep learning, weakly supervised deep learning, and deep reinforcement learning [45]. Figure 1.6 illustrates the distinct categories of deep learning models and provides some examples of the state-of-the-art model in each category. The following subsections provide a concise description of these categories along with the primary algorithms belonging to each category [23]. This section will no hesitation serve up as a groundwork to the key subject matters of the later chapters of this book.

1.5.1 Deep Supervised Learning

Supervised learning is considered by the way as the most popular learning paradigm for either machine or deep learning algorithms. As the name "supervised" imply, the learning process is performed to map input data to known targets (data labels), where the set of trainable parameters of the deep learning model are tuned based on the model's inputs to better fit the targeted outputs. In other words, in supervised learning, the training data supplied to the model act as the supervisor that educates the model to estimate the output accurately. It simulates the idea of school learning where a student learns under the supervision of the instructor. Hence, the main

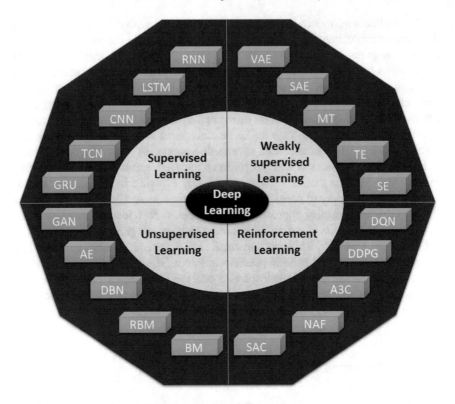

Fig. 1.6 Taxonomy of current deep learning approaches

objective of supervised learning is to optimize the model parameters in such a way that minimizes the difference between the model's estimated outputs and the actual data labels provided by humans [46].

To land this definition into a tangible example, consider considering operating in an IoT system, in which a particular transaction responsible for detecting whether a specific sample of IoT traffic would be normal or abnormal. This remark, "normal" or "abnormal", can be considered as the class label. Where the input features might contain packet information, transmission time, reception time, arrival rate, etc. Thanks to providing the model with a dataset containing only annotated samples, in which each sample has its corresponding ground-truth label always being determined by human experts (supervisors), hence the concept of supervision came into play. Generally, the majority of applications of deep learning in IoT-based systems that are receiving attention is belonging to this learning paradigm, such as Traffic Speed Forecasting, Load forecasting, activity recognition, gait recognition, etc. Figure 1.7 display the workflow of the supervised deep learning model [47].

At the training stage, the training samples and the corresponding labels are fed into deep learning models, after each training iteration, the difference between model output and the data labels are measured, and the accordingly model parameters

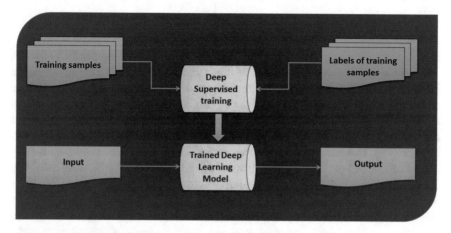

Fig. 1.7 Visual representation of workflow of supervised Deep Learning

are updated till the model converges. Finally, at the inference stage, the previously unused inputs are passed to the learned model to estimate corresponding outputs, the difference between the estimated and the actual label is measured to evaluate the performance of the model.

Furthermore, the supervised deep learning approaches can be classified into two subcategories according to the continuity of data labels. First, deep supervised learning with continuous output values and data labels are referred to as regression, while on the other hand, deep supervised learning with discrete output values and data labels, is referred to as classification.

1.5.1.1   Regression

Regression is a popular category of supervised learning for predicting continuous output-based current values of input data. This is also referred to as predictive analysis or forecasting. Specifically, the deep learning model receives labeled data consisting of the number of explanatory variables or features and the corresponding continuous target variable, then the model is trained to model the relationships between them to be able to accurately predict the target output for new input samples. Predicting the electrical load in a smart grid system can be considered as a typical example for a regression task. In regression analysis can be implemented in two forms. First, scalar regression, which refers to the situations where the predicted target is a continuous scalar value. Second, vector regression, which refers to the situations where the predicted target is a set of continuous values under vectorized format [48].

1.5.1.2   Classification

It is a category of supervised learning, in which a deep learning model is trained to learn the inherent representations of input representations with the main aim to identify and predict the categorical class of novel inputs. These targeted classes are an unordered and discrete set of values that can be interested as the group membership of the data samples. In this regard, the classification problems can be further divided into four subcategories according to the number of class labels involved in the data [46]. This includes binary classification, one-class classification, multiclass classification, and multilabel Classification. First, the binary classification considers the situation where the input data belongs to only two classes. For example, given an IoT-based system in which the service providers want to detect anomalies in client's data [49]. Hence, the deep learning model is trained using samples that are annotated either as normal or as an anomaly. Second, one-class classification seeks to train a deep learning model using only a single class of data. This is especially valuable in an imbalanced classification dataset with an over-abundance of data of a certain class. Consider the previous example where the training set contains millions of normal samples and just hundreds of anomaly samples. Third, multiclass classification considers the classification problem that contains three or more classes. A typical example for multiclass classification is cyber-attack classification. Given an IoT system that is known to be vulnerable to four categories of cyber-attacks i.e., Ransomware, spyware, Phishing, and eavesdropping. The identification of the category of underlying attack is essential to determine the proper countermeasures, or recovery procedures. Hence, deep learning model is trained using the labeled IoT traffic flows as an input to determine the attack class of input samples. Finally, multi-label classification is considered as a category of classification similar to multiclass classification, in which more than one class label need to be predicted, which implies that the class labels are mutually non-exclusive. In multi-label classification, one or more labels can be jointly considered as target output for each input sample, and the model should predict this target jointly. For example, face recognition as an example, where single image contains multiple persons each with different class label.

In the spectrum of supervised learning, Convolutional networks recurrent networks are the most common deep learning models. Convolutional networks include Dense convolution Network (DenseNet) [50], residual convolution network (ResNet) [51], Temporal Convolution Network (TCN) [52], etc. On the other hand, recurrent networks include Gated Recurrent Unit (GRU), Long Short-term Memory (LSTM) [38], and simple recurrent neural network (simpleRNN).

1.5.2 Deep Unsupervised Learning

Unsupervised Learning is another common learning paradigm for deep learning models, which alleviates the need for supervision during the model training by

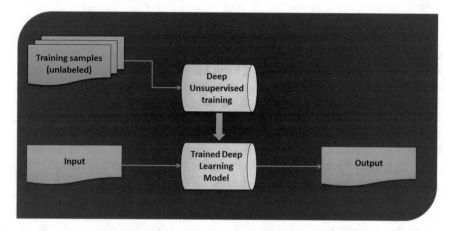

Fig. 1.8 Visual representation of workflow of unsupervised deep learning

allowing the model to be using its own to uncover hidden patterns and learn important data representations from unlabeled data [53]. The main goal of unsupervised learning is to capture high-order interrelationships from the noticed or discernible data for synthetic or analytical reasons under the absence of information about target classes. The potentials of deep unsupervised learning to discover relationships and variations in information render it a perfect solution for exploratory data analysis, cross-selling policies, client grouping, and similar complex tasks [54]. Figure 1.8 display the workflow of the unsupervised deep learning model.

1.5.2.1 Clustering

Clustering is a common subfield of unsupervised learning, in which the deep learning model seeks to learn an inherent representation of data to organize them into distinct and momentous clusters without having any form of supervision or knowledge about cluster membership. Every single cluster defines a group of samples that are having a high degree of interrelation and are different from the samples of other clusters. Hence, clustering is a popular big data analysis tool for extracting valuable correlation information necessary for structuring data into distinct groups. With the vast amount of daily generated data in the IoT-based system, deep learning-based clustering become a suitable choice to provide useful insights about clients for business owners [47].

1.5.2.2 Dimensionality Reduction

Dimensionality reduction is defined as popular unsupervised learning that aims to lower the input dimension by reducing the number of constituent input variables to avoid the curse of dimensionality. The curse of dimensionality refers to a situation where the input samples consist of a large number of variables which imply large storage space and increase the computational complexity of machine learning algorithms limiting the predictive modeling capabilities [55]. Thus, unsupervised dimensionality reduction is essential to act as an essential preparation step of any learning process to take the responsibility of data cleaning, noise removal, redundancies elimination, and transforming the data representation into simpler or more accessible than the original representations. The concept of *simpler* representation can be defined in different ways including lower-dimensional representations, sparse representations, and independent representations [35]. Fortunately, the resource-constrained nature of IoT devices makes dimensionality reduction is essential to deploy intelligent solutions in the IoT environment. Moreover, AI-based security solutions can benefit from dimensionality reduction to make the model emphasize the most relevant features necessary to improve the efficiency of AI model and the depending on AI solution.

1.5.2.3 Self-Supervised Learning

Self-supervised learning is viewed as an emerging field of unsupervised learning since there is no manual label involved. Nevertheless, intently speaking, unsupervised learning focuses on discovering certain patterns of data, as with community discovery and clustering, while self-supervised learning emphasis recovering, which is considered as a part of supervised configurations [56].

The concept of self-supervised learning is firstly presented in the field of robotics, in which the training data is routinely annotated by leveraging the relationships between various sensory inputs. Later, the artificial intelligence community studied and advanced this concept. In the AAAI 2020, Yann LeCun (The Turing award winner) redefined the concept of self-supervised learning to be "the machine predicts any parts of its input for any observed part". Merging the legacy definition and LeCun's definition, supervised learning can be defined as emerging n emerging unsupervised representation learning solution that enables automated data labeling leverages input data itself as supervision while eliminating human supervision. In other words, It enables the deep learning models to trains themselves by leveraging one portion of the data to estimate and foresee the other portion and accurately generate labels, hence enables shifting from unsupervised settings to a supervised one that can benefit a variety of downstream tasks [57]. According to the continuously evolving nature of IoT-based systems, with a large amount of unlabeled data, self-supervised learning is foreseen to offer a promising solution for different IoT applications especially for security-related applications such as intrusion detection, attack defending, etc.

In this spectrum of unsupervised learning, autoencoders (AEs), generative adversarial networks (GAN), deep belief networks (DBN), self-organizing map (SoM), Boltzmann machine (BM), Restricted Boltzmann machine (RBM) Transformer models (i.e., BERT), and transformation equivariant models are all examples of deep learning models.

1.5.3 Deep Weakly Supervised Learning

The weakly supervised learning is another learning paradigm that lies between supervised and unsupervised learning, in which a substantial quantity of data supervision is not available [58]. This includes 1) incomplete supervision, where only an insignificant portion of data was labeled, such as semi-supervised learning, Active Learning (AL), and domain adaptation (DA); 2) inexact supervision, in which only coarsegrained labels are assigned, but not as precise as real ground truth, and 3) inaccurate supervision, which refer to a situation where data are assigned some erroneous labels that are not permanently ground truth [59].

In incomplete supervision, active learning can be employed by including a domain expert to obtain data labels for unlabeled samples upon request. it believes that the expense of labeling operations mainly hinges on the total of queries, hence the objective is to reduce the number of queries as possible [60]. In contrast, deep semi-supervised learning can be regarded as a learning paradigm that combines the advantages of both supervised learning and unsupervised learning by allowing the deep learning model to be trained using both unlabeled and labeled samples, without any human interventions [61, 62]. Figure 1.9 display the workflow of the semi-supervised deep learning model. The notion behind this is that unlabeled samples offer valuable clues on the way the data are normally distributed in the space while exploring this distribution is beneficial for improving the performance of a model trained with partial supervision of labeled data. For instance, a forceful model should make steady and smooth predictions under arbitrary transformations alongside the data manifold or keep away from setting its decision frontier on excessive intensity zones of data distribution. This permits the model to infer important patterns and discover relationships between the dataset and the target variable. In this regard, a

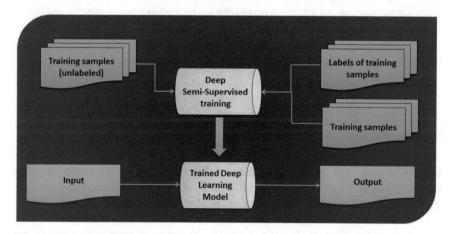

Fig. 1.9 Visual representation of workflow of semi-supervised deep learning

Table 1.2 Comparison between different learning paradigms

Features	Supervised learning	Unsupervised learning	Weakly supervised
Process	The Deep Learning model, input data, and its label are given	Deep learning model, only input data is given	Deep Learning model, labeled input, and unlabeled input data are given
Input data	Labeled data	Unlabeled	Partially labels Inexact labels Inaccurate label
Categories	Regression classification	Clustering Dimensionality reduction	Domain adaption Active learning
Computational complexity	A simpler method	Computationally complex	More complex than supervised
Use of data	Learn to map between the input and the targets	Learn inherent patterns of data	Learn to map between the input and the targets, and data distribution
Accuracy	Highly efficient and dependable method	Less accurate and trustworthy method	More efficient than unsupervised
Real-time	Offline learning	Real-time learning	Offline learning
No. of classes	Known	Unknown	Known
Proximity to real intelligence	Not close	Closer to real	Closer to real than supervised learning

variety of models can be employed as semi-supervised learners including teacher-student models (i.e., temporal-ensembling, Mean-Teacher, π-model), domain adaption models, semi-supervised generative models, and semi-supervised autoencoders. Also, meta-learning can be combined under the spectrum of semi-supervised where a meta-model can be trained to distill the knowledge of the way to update models including few samples. Table 1.2 compares supervised learning with unsupervised, and semi-supervised learning based on a different set of features.

1.5.4   Deep Reinforcement Learning

Reinforcement Learning (RL) provides a framework for agents to get involved in the environment and automatically make a decision based on the perceived states or conditions. This environment is not restricted to the lab environment but also expands to reach out to a real-world IoT environment. These agents are not essentially a software thing i.e., IoT software, routing software, etc. As an alternative, they can be embedded into hardware components of robots, autonomous vehicles. An embedded agent is possibly the superlative method to abundantly apply and evaluate reinforcement learning as the physical things engage in a variety of interactions with

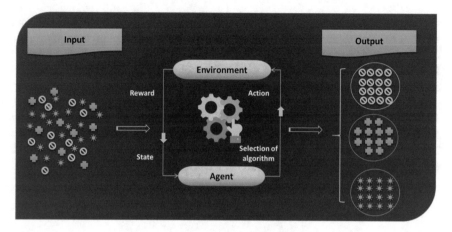

Fig. 1.10 Visual representation of workflow of Deep Reinforcement Learning

different aspects of the real world and perceive responses. Figure 1.10 display the workflow of the reinforcement deep learning model [63].

Generally, any environment possesses a different set of states that can be moderately or completely noticeable by the agent placed into this environment. The agent usually performs a variety of actions through which it interacts with the environment and changes its current state to a new state. Following the change of the state, the agent receives either a reward or punishment value according to the nature of the change. Hence, the agent applies a specific learning policy to manage and determine the most suitable action to perform under a certain state with the main objective to maximize the achieved forthcoming reward [64].

In other words, reinforcement learning strongly imitates human psychology for experiencing the real-world environment. Where incorrect human behaviors often lead to a convincing form of loss or negative consequence which necessitate evading them in the future, whereas the right human activities result in some benefit or rewards and necessitate encouraging them into the future. Reinforcement learning has been demonstrating its success in different applications thought-out the years. However, standard Reinforcement learning suffers from scalability limitations, especially in complex and wide-scale environments. Conventional reinforcement learning is prone to the previously mentioned phenomena of the "curse of dimensionality". This is where Deep Learning came into existence to address this limitation paving the way for a new field known as Deep Reinforcement Learning (DRL), which is an integration of RL and deep network [64, 65]. Unlike the before mentioned categories of deep learning, the DRL has the ability to perform dynamic reactions for security changes in the IoT environment. For instance, given a trained supervised, unsupervised, or weakly supervised model and deployed for attack detection; when an unseen and continually changing attack is presented to the system, the model almost will fail to identify as it needs to learn the pattern of those attacks from a new simulated dataset. On the other hand, The DRL can effectively supervise instantaneous dynamic environments (i.e., IoT). It performs the learning process on the basis of trial and error in the environment, hence reduces the effort related to the simulation of a complicated IoT environment. The distinct characteristics of Deep Reinforcement Learning qualify it as a promising dataset-free solution for addressing security and forensic challenges of IoT-system [63].

In this spectrum of Deep Reinforcement Learning, Deep Q Network (DQN), Deep Deterministic Policy Gradient (DDPG), Asynchronous Advantage Actor-Critic Algorithm (A3C), Q-Learning with Normalized Advantage Functions (NAF), Soft Actor-Critic (SAC), are all examples of Reinforcement Learning algorithms. A detailed explanation of the role of DRL in securing the IoT-baed system is discussed in a later chapter of this book.

1.6   Privacy in Internet of Things

The word "privacy" originates from the old French term "privaut´e", which implies confidentiality, mystery, secrecy, or isolation. The concept of Privacy has an extended history of definitions and implications. The development of contemporary awareness and innovative culture causes a progressive change in the implication of "privacy". By 1590, it intended a confidential issue or concealment and was like the "seclusion" in the sixteenth century. Then, it was defined as a "state of freedom from intrusion" after 1814. By 1890, Warren et al. [66] released the paper "The Right to Privacy", obviously advising the conception of privacy. At the present time, the concept of privacy means the skill of persons or companies to selectively isolate and communicate their information, specifically for persons. Consequently, the notion of privacy is profoundly entrenched in human lives. Nevertheless, with the advent of big data, leakages and/or violations of privacy takes place constantly and worldwide. In 2018, Verizon's report announced that the amount of data leaked per day exceed seven million slices of data [67]. As an example, the Facebook Cambridge Analytica data scandal was undoubtedly one of the biggest Internet events in 2018. The information of more than 87 million Facebook clients was released by the question-and-answer application with no permission and warning. In the USA, the private information of security staff and the American military was exposed via a fitness application called ("Polar") [68, 69]. In India, the biometric identification system ("Aadhaar") directed by the government is prone to malevolent threats owing to the absence of access control procedures, letting the personal information of residents simply reachable [70]. Hence, privacy is a problem of interest for almost all technologies for ages. The complex and dynamic structure of IoT-based systems enables the integration of a wide range of different technologies, which imposes a wide variety of privacy challenges.

One of the major motives behind the privacy interest in IoT environments is the thriving of deep learning, which enables developing models to learn from IoT data to offer intelligent solutions that can enhance the functionality or the security of the underlying system. Besides, the design of an IoT-based system typically encounters plenty of data interchange and pervasive deep learning solutions, and the IoT data is likely to entail private information and sensitive details that are endangered at either the device or communication stage. By this moment, the massive data exchange needed by deep learning in IoT-based system would impose a key challenge for protecting the privacy of individuals, which has been gaining increasing interest from academia, industry, and governments. In view of this, rigorous regulations such as the European Commission's General Data Protection Regulation (GDPR) [71] (https:// gdpr.eu/data-privacy/), and the US Consumer Privacy Bill of Rights [72] (https:// www.congress.gov/bill/116th-congress/senate-bill/2968/text) have been devised to safeguard the privacy of individuals. As an example, data minimalization law (Article 5 of GDPR) and the consent (Article 6 of GDPR) constrain the data aggregation and storing to only what is user-permitted and completely essential for processing. The smartphone manufacturers i.e., Samsung and Apple Inc., employ some Differential Privacy (DP) techniques to keep the privacy of their customers. By 2016, Scholars have presented the paradigm of Federated Learning (FL) [73] to enable mobile and geographically distributed devices to collaboratively train ML or DL using their own private local data. In this way, the FL alleviates the need to share or send the data to a centralized as the data remain kept locally, thereby the privacy of data can be easily kept and supervised.

In the field of IoT, the individuals' sensory measurements and even emotive information could be aggregated and passed to the deep learning model, where different kinds of privacy assurance can be employed, which can negatively influence the efficiency of the model. this phenomenon is known as "privacy-efficiency trade-off" or "privacy-accuracy trade-off". Thus, ensuring better equilibrium between the efficiency of the model and the data privacy is an essential requirement for any deep learnings' solution [74, 75]. The role of deep learning for privacy can have two sides, one as a preserver, and the other as a violator. As with other technologies, individuals can use deep learning in such a way that preserves IoT privacy or violate it either intentionally or unintentionally [76].

With recent advancements in IoT communications, the privacy challenges largely varied. First, individuals have given an interest in the value of data from different perspectives (i.e., data mining), and at present, a large amount of data is centered under the control of large enterprises. The IoT seeks to alleviate the monopoly of data and build an expansive data sharing and exchanging. Enormous confidential data and decreased complexity in attaining data will impose greater obligations on privacy preservation techniques, and simultaneously instigate supervisory complexities and even prompt violations against data. Secondly, more intelligent solutions operating on network edges and mobile devices, rising the danger of data leakage or privacy violations. Preserving privacy in view of these resource-limited environments is an extremely challenging target [44]. After All, the equilibrium between privacy-preservation and the efficiency of intelligent solutions (i.e., services, applications) also deserves to gain much interest. Therefore, IoT necessitates paying consideration toward data proprietorship, access privileges, applicable control, acceptable regulations, and privacy-efficiency trade-off. Privacy preservation is considered a key enabler for reliable and trustworthy IoT systems in the future. Therefore, deep learning and privacy-preservations are closely related paradigms that require much investigation and analysis [77].

In a later chapter of this book, a comprehensive description is provided for the role of deep learning in preserving the privacy of IoT-based system in such a way the help the beginners and researchers to clear visions and understanding of this subject in order to further enable the design of privacy-preserving IoT solutions.

1.7   Summary and Learnt Lessons

The detailed discussions in different sections of this chapter can provide us with a set of important learned lessons that can be briefly described as follow:

- The emergence of IoT increases the security vulnerabilities, breaches, attacks at different elements of IoT-based systems. The concept of IoT security differs from traditional cybersecurity. Hence, IoT security is a challenging and essential aspect for assuring the reliability of IoT-based systems.
- The integration of digital forensics into IoT brings us with the concept of IoT forensics that mainly concerns investigating the digital evidence in IoT-based systems.
- Artificial Intelligence is an evolving discipline of computer science that has achieved great contribution in protecting IoT security using machine learning paradigms. However, the limited capabilities of ML algorithms make them unable to satisfy the security and forensic requirements of IoT-based systems.
- Deep learning emerged as a subfield of ML designed on basis of neural networks and is viewed as a promising solution for challenging problems of IoT security and/or forensics.
- Privacy preservation is another cornerstone of IoT, which gain much interest from academia, and industry. Deep learning is considered a two-sided paradigm that can be used to protect privacy or violate it.

1.8   Book Organization

What This Book Covers?

To answer this question, the organization of chapters in this book, along with the topics they cover is briefly discussed as follow:

This chapter, *Conceptualization of Security, Forensics, and Privacy of Internet of Things: An Artificial Intelligence Perspective,* discuss the main concepts behind the realization of reliable internet IoT including IoT security, IoT forensics, and IoT privacy. It also, discuss the potential of Artificial intelligence, machine learning, deep learning for securing an IoT system The derivation of deep learning and the taxonomy of classifying deep learning models is also discussed in this chapter.

Chapter 2, *Internet of Things, Preliminaries and Foundations*, discusses the definition of IoT system, the layered architecture of IoT system, and provide a fine-grained definition for layers of IoT system, along with their responsibilities, elements, and ranking. The chapter also argues important concepts and computing paradigms greatly related to the IoT system, such as cloud computing, fog computing, and edge computing.

Chapter 3, *Internet of Things Security Requirements, Threats, Countermeasures*, provide a thorough discussion about different attack surfaces and the potential vulnerabilities in IoT systems. This entails different threats and attacks across different tiers or layers of IoT systems. Then, a taxonomy for dividing the attack surfaces into multiple categories is presented along with the primary security threats and possible vulnerabilities per each category. Furthermore, a novel attack surfaces triggered by the mutually dependent, interrelated, and co-operating environments of IoT systems.

Chapter 4, *Digital Forensics in Internet of Things*, presents a compact discussion of the theoretical approaches, essential challenges, and research developments of digital Forensics in IoT-based system. Besides, it figures out the requirement for standardization the forensics procedures, as this is considered as a crucial phase towards high-level quality cross-jurisdictional forensics articles and ideal procedures for assuring cyber-physical security.

Chapter 5, *Supervised Deep Learning for Secure Internet of Things*, present an in-depth discussion for the supervised deep learning approaches for maintaining the security of IoT system. This includes the state-of-the-art supervised algorithms for securing IoT systems, the main characteristics, corresponding advantages and disadvantages, the corresponding role in IoT security. Also, provide detailed tabular comparisons and figures that indicate help the readers to grasp the learned lessons.

Chapter 6, *Unsupervised Deep Learning for Secure Internet of Things*, present a comprehensive discussion for the learning from unlabeled IoT data using unsupervised deep learning approaches for the purpose of securing the IoT environments. This involves the state-of-the-art unsupervised algorithms along with their main characteristics, the corresponding advantages and disadvantages, some use cases in IoT security. Also, provide a detailed tabular comparisons and figures that help the readers obtain full understanding of the methodology of unsupervised learning.

Chapter 7, *Semi-supervised Deep Learning for Secure Internet of Things*, present a detailed discussion about learning from partially labeled data using the semi-supervised deep learning approaches for maintaining the security of IoT system. This includes the state-of-the-art supervised algorithms for securing IoT systems, the main characteristics, corresponding advantages and disadvantages, the corresponding role in IoT security. Furthermore, provide a detailed tabular comparisons and figures that aims to help the readers to grasp the learned lessons.

Chapter 8, *Reinforcement Learning for Secure Internet of Things*, introduce and discuss the potential of deep reinforcement learning for maintaining the security of IoT system. This includes the reinforcement learning supervised algorithms, the key similarities, and differences, the related advantages, and disadvantages, the condition of usages securing IoT environment. Moreover, illustrative diagrams are also presented to emulate the workflow of different reinforcement learning frameworks to assist the readers to completely comprehend the role of reinforcement learning in securing the IoT environments.

Chapter 9, *Federated Learning for Privacy-Preserving Internet of Things*, discuss privacy considerations of applying intelligent solutions in IoT-based system. The federated learning paradigm is extensively discussed, categorized, and analyzed for privacy preservation. Moreover, it also deliberates the role of differential-privacy, multi-parity computing, and other methods of privacy preservations.

Chapter 10, *Challenges, Opportunities, and Future Prospects*, introduce the current and most recent challenges facing the deep learning community when designing privacy and security preserving IoT system, also present possible solutions that could be investigated by the academic or industrial community in near future. Furthermore, the convergence of deep learning approaches and IoT technologies and still in its early phases paving the way toward more reliable and trustworthy IoT solutions which are figured out as a part of future work in either academia or marketspace.

References

1. F. Javed, M.K. Afzal, M. Sharif, B.S. Kim, Internet of things (IoT) operating systems support, networking technologies, applications, and challenges: a comparative review. IEEE Commun. Surv. Tutorials (2018). https://doi.org/10.1109/COMST.2018.2817685
2. B. Russell, D. Van Duren, *Practical Internet of Things Security* (2016)
3. M.A. Al-Garadi, A. Mohamed, A.K. Al-Ali, X. Du, I. Ali, M. Guizani, A survey of machine and deep learning methods for internet of things (IoT) security. IEEE Commun. Surv. Tutorials (2020). https://doi.org/10.1109/COMST.2020.2988293
4. N. Neshenko, E. Bou-Harb, J. Crichigno, G. Kaddoum, N. Ghani, Demystifying IoT security: an exhaustive survey on IoT vulnerabilities and a first empirical look on internet-scale IoT exploitations. IEEE Commun. Surv. Tutorials (2019). https://doi.org/10.1109/COMST.2019.2910750
5. F. Hussain, R. Hussain, S.A. Hassan, E. Hossain, Machine learning in IoT security: current solutions and future challenges. IEEE Commun. Surv. Tutorials (2020). https://doi.org/10.1109/COMST.2020.2986444
6. F.O. Olowononi, D.B. Rawat, C. Liu, Resilient machine learning for networked cyber physical systems: a survey for machine learning security to securing machine learning for CPS. IEEE Commun. Surv. Tutorials (2021). https://doi.org/10.1109/COMST.2020.3036778
7. M. Zolanvari, M.A. Teixeira, L. Gupta, K.M. Khan, R. Jain, Machine learning-based network vulnerability analysis of industrial internet of things. IEEE Internet Things J. (2019). https://doi.org/10.1109/JIOT.2019.2912022
8. A. Al-Fuqaha, M. Guizani, M. Mohammadi, M. Aledhari, M. Ayyash, Internet of things: a survey on enabling technologies, protocols, and applications. IEEE Commun. Surv. Tutorials (2015). https://doi.org/10.1109/COMST.2015.2444095
9. H. Gill, *NSF Perspective and Status on Cyber-Physical Systems* (2006). Austin. Internet http//varma.ece.C.edu/CPS/Presentations/gill.pdf. [zuletzt aufgesucht am 1.04. 2015]
10. A. Humayed, J. Lin, F. Li, B. Luo, Cyber-physical systems security—a survey. IEEE Internet Things J. (2017). https://doi.org/10.1109/JIOT.2017.2703172
11. S. Li, Q. Ni, Y. Sun, G. Min, S. Al-Rubaye, Energy-efficient resource allocation for industrial cyber-physical IoT systems in 5G era. IEEE Trans. Ind. Inform. (2018). https://doi.org/10.1109/TII.2018.2799177

12. S. Li, K.K.R. Choo, Q. Sun, W.J. Buchanan, J. Cao, IoT forensics: Amazon echo as a use case. IEEE Internet Things J. (2019). https://doi.org/10.1109/JIOT.2019.2906946
13. J. Hou, Y. Li, J. Yu, W. Shi, A survey on digital forensics in internet of things. IEEE Internet Things J. (2020). https://doi.org/10.1109/JIOT.2019.2940713
14. M. Stoyanova, Y. Nikoloudakis, S. Panagiotakis, E. Pallis, E.K. Markakis, A survey on the internet of things (IoT) forensics: challenges, approaches, and open issues. IEEE Commun. Surv. Tutorials (2020). https://doi.org/10.1109/COMST.2019.2962586
15. F. Ding, G. Zhu, M. Alazab, X. Li, K. Yu, Deep-learning-empowered digital forensics for edge consumer electronics in 5G HetNets. IEEE Consum. Electron. Mag. (2020). https://doi.org/10.1109/MCE.2020.3047606
16. N. Koroniotis, N. Moustafa, E. Sitnikova, Forensics and deep learning mechanisms for botnets in internet of things: a survey of challenges and solutions. IEEE Access. (2019). https://doi.org/10.1109/ACCESS.2019.2916717
17. The Cambridge handbook of artificial intelligence. Choice Rev. Online (2015). https://doi.org/10.5860/choice.187061
18. J. Zhang, D. Tao, Empowering things with intelligence: a survey of the progress, challenges, and opportunities in artificial intelligence of things. IEEE Internet Things J. (2021). https://doi.org/10.1109/JIOT.2020.3039359
19. S. Singh, P.K. Sharma, B. Yoon, M. Shojafar, G.H. Cho, I.H. Ra, Convergence of blockchain and artificial intelligence in IoT network for the sustainable smart city. Sustain. Cities Soc. (2020). https://doi.org/10.1016/j.scs.2020.102364
20. A.L. Samuel, Some studies in machine learning using the game of checkers. IBM J. Res. Dev. (1959)
21. T.M. Mitchell, Does machine learning really work? AI Mag. (1997)
22. J. Wang, C. Jiang, H. Zhang, Y. Ren, K.C. Chen, L. Hanzo, Thirty years of machine learning: the road to pareto-optimal wireless networks. IEEE Commun. Surv. Tutorials (2020). https://doi.org/10.1109/COMST.2020.2965856
23. S. Raschka, V. Mirjalili, *Python Machine Learning: Machine Learning & Deep Learning with Python, Scikit-Learn and TensorFlow 2*, 3rd edn. (2019)
24. B.H. Yang, H. Asada, Progressive learning and its application to robot impedance learning. IEEE Trans. Neural Netw. (1996). https://doi.org/10.1109/72.508937
25. L. Liu, X. Xu, Y. Liu, Z. Ma, J. Peng, A detection framework against CPMA attack based on trust evaluation and machine learning in IoT network. IEEE Internet Things J. (2021). https://doi.org/10.1109/JIOT.2020.3047642
26. M. Shafiq, Z. Tian, A.K. Bashir, X. Du, M. Guizani, CorrAUC: a malicious Bot-IoT traffic detection method in IoT network using machine-learning techniques. IEEE Internet Things J. (2021). https://doi.org/10.1109/JIOT.2020.3002255
27. F. Rosenblatt, The perceptron: a probabilistic model for information storage and organization in the brain. Psychol. Rev. (1958). https://doi.org/10.1037/h0042519
28. S. Narayan, The generalized sigmoid activation function: competitive supervised learning. Inf. Sci. (Ny) (1997). https://doi.org/10.1016/S0020-0255(96)00200-9
29. C. Nwankpa, W. Ijomah, A. Gachagan, S. Marshall, *Activation Functions: Comparison of trends in Practice and Research for Deep Learning* (Nov. 2018), [Online]. Available: http://arxiv.org/abs/1811.03378
30. I. Goodfellow, Y. Benigo, A. Courville, Deep Learning (Adaptive Computation and Machine Learning series): 9780262035613 (Amazon.com: Books, MIT Press, 2016)
31. A.L. Maas, A.Y. Hannun, A.Y. Ng, Leaky ReLU. ICML Work. Deep Learn. Audio, Speech Lang. Process. (2013)
32. A.L. Maas, A.Y. Hannun, A.Y. Ng, *Rectifier Nonlinearities Improve Neural Network Acoustic Models* (2013)
33. D.A. Clevert, T. Unterthiner, S. Hochreiter, *Fast and Accurate Deep Network Learning by Exponential Linear Units (ELUs)* (2016)
34. P. Ramachandran, B. Zoph, Q.V. Le, Swish A Self-Gated Activation Function. arXiv (2017)
35. I. Goodfellow, Y. Bengio, A. Courville, Deep Learning (Google Books, MIT Press, 2016)

36. C. Dugas, Y. Bengio, F. Bélisle, C. Nadeau, R. Garcia, *Incorporating Second-Order Functional Knowledge for Better Option Pricing* (2001)
37. J. Turian, J. Bergstra, Y. Bengio, *Quadratic Features and Deep Architectures for Chunking* (2009). https://doi.org/10.3115/1620853.1620921
38. G. Klambauer, T. Unterthiner, A. Mayr, S. Hochreiter, *Self-Normalizing Neural Networks* (2017)
39. D. Hendrycks, K. Gimpel, *Gaussian Error Linear Units (GELUs)* (Jun. 2016), [Online]. Available: http://arxiv.org/abs/1606.08415
40. J.T. Barron, *Continuously Differentiable Exponential Linear Units* (Apr. 2017), [Online]. Available: http://arxiv.org/abs/1704.07483
41. A.G. Ivakhnenko, V.G. Lapa, *Cybernetic Predicting Devices* (CCM Information Corporation, 1965)
42. D. Zou, Y. Cao, D. Zhou, Q. Gu, Gradient descent optimizes over-parameterized deep ReLU networks. Mach. Learn. (2020). https://doi.org/10.1007/s10994-019-05839-6
43. F. Chollet, *Deep Learning with Python, Manning* (2018)
44. A. Alwarafy, K.A. Al-Thelaya, M. Abdallah, J. Schneider, M. Hamdi, A Survey on security and privacy issues in edge-computing-assisted internet of things. IEEE Internet Things J. (2021). https://doi.org/10.1109/JIOT.2020.3015432
45. A. Zhang, Z.C. Lipton, M. Li, A.J. Smola, *Dive into Deep Learning* (2020). URL: https://d2l.ai/
46. Q. Zhang, L.T. Yang, Z. Chen, P. Li, A survey on deep learning for big data. Inf. Fusion (2018). https://doi.org/10.1016/j.inffus.2017.10.006
47. M. Mohammadi, A. Al-Fuqaha, S. Sorour, M. Guizani, Deep learning for IoT big data and streaming analytics: a survey. IEEE Commun. Surv. Tutorials (2018). https://doi.org/10.1109/COMST.2018.2844341
48. S. Lathuiliere, P. Mesejo, X. Alameda-Pineda, R. Horaud, A comprehensive analysis of deep regression. IEEE Trans. Pattern Anal. Mach. Intell. (2020). https://doi.org/10.1109/TPAMI.2019.2910523
49. A. Aldweesh, A. Derhab, A.Z. Emam, Deep learning approaches for anomaly-based intrusion detection systems: a survey, taxonomy, and open issues. Knowledge-Based Syst. (2020). https://doi.org/10.1016/j.knosys.2019.105124
50. G. Huang, Z. Liu, L. Van Der Maaten, K.Q. Weinberger, *Densely Connected Convolutional Networks* (2017). https://doi.org/10.1109/CVPR.2017.243
51. K. He, X. Zhang, S. Ren, J. Sun, *Deep Residual Learning for Image Recognition* (2016). https://doi.org/10.1109/CVPR.2016.90
52. A. Herle, J. Channegowda, D. Prabhu, *A Temporal Convolution Network Approach to State-of-Charge Estimation in Li-ion Batteries* (2020). https://doi.org/10.1109/INDICON49873.2020.9342315
53. G. Wilson, D.J. Cook, A survey of unsupervised deep domain adaptation. ACM Trans. Intell. Syst. Technol. (2020). https://doi.org/10.1145/3400066
54. G.-J. Qi, J. Luo, Small data challenges in big data era: a survey of recent progress on unsupervised and semi-supervised methods. IEEE Trans. Pattern Anal. Mach. Intell. (2020). https://doi.org/10.1109/tpami.2020.3031898
55. L. Xu, C. Jiang, Y. Ren, H.H. Chen, Microblog dimensionality reduction-a deep learning approach. IEEE Trans. Knowl. Data Eng. (2016). https://doi.org/10.1109/TKDE.2016.2540639
56. L. Jing, Y. Tian, Self-supervised visual feature learning with deep neural networks: a survey. IEEE Trans. Pattern Anal. Mach. Intell. (2020). https://doi.org/10.1109/tpami.2020.2992393
57. S. Bucci, A. D'Innocente, Y. Liao, F.M. Carlucci, B. Caputo, T. Tommasi, Self-supervised learning across domains. IEEE Trans. Pattern Anal. Mach. Intell. (2021). https://doi.org/10.1109/TPAMI.2021.3070791
58. Y.F. Li, L.Z. Guo, Z.H. Zhou, Towards safe weakly supervised learning. IEEE Trans. Pattern Anal. Mach. Intell. (2021). https://doi.org/10.1109/TPAMI.2019.2922396
59. D. Zhang, J. Han, G. Cheng, M.H. Yang, Weakly supervised object localization and detection: a survey. IEEE Trans. Pattern Anal. Mach. Intell. (2021). https://doi.org/10.1109/TPAMI.2021.3074313

60. J. Bernard, M. Hutter, M. Zeppelzauer, D. Fellner, M. Sedlmair, Comparing visual-interactive labeling with active learning: an experimental study. IEEE Trans. Vis. Comput. Graph. (2018). https://doi.org/10.1109/TVCG.2017.2744818
61. X. Wang, Y. Chen, W. Zhu, *A Comprehensive Survey on Curriculum Learning* (PAMI, 2020)
62. X. Wang, Y. Chen, W. Zhu, A survey on curriculum learning. IEEE Trans. Pattern Anal. Mach. Intell. (2021). https://doi.org/10.1109/TPAMI.2021.3069908
63. N.C. Luong et al., Applications of deep reinforcement learning in communications and networking: a survey. IEEE Commun. Surv. Tutorials (2019). https://doi.org/10.1109/COMST.2019.2916583
64. B.R. Kiran et al., Deep reinforcement learning for autonomous driving: a survey. IEEE Trans. Intell. Transp. Syst. (2021). https://doi.org/10.1109/TITS.2021.3054625
65. L. Lei, Y. Tan, K. Zheng, S. Liu, K. Zhang, X. Shen, Deep reinforcement learning for autonomous internet of things: model, applications and challenges. IEEE Commun. Surv. Tutorials (2020). https://doi.org/10.1109/COMST.2020.2988367
66. S. Warren, L. Brandeis, The right to privacy 1890. Harv. Law Rev. (1890)
67. A. Datoo, Data in the post-GDPR world. Comput. Fraud Secur. (2018). https://doi.org/10.1016/S1361-3723(18)30088-5
68. S. Boukoros, M. Humbert, S. Katzenbeisser, C. Troncoso, *On (the lack of) Location Privacy in Crowdsourcing Applications* (2019)
69. Y. Sun, J. Liu, J. Wang, Y. Cao, N. Kato, When machine learning meets privacy in 6G: a survey. IEEE Commun. Surv. Tutorials (2020). https://doi.org/10.1109/COMST.2020.3011561
70. A.K. Jain, K. Nandakumar, Biometric authentication: system security and user privacy. Computer (Long. Beach. Calif.) (2012). https://doi.org/10.1109/MC.2012.364
71. B. Custers, A.M. Sears, T. Tani, D.K. Mulligan, *EU Personal Data Protection in Policy and Practice* (2019)
72. B.M. Gaff, H.E. Sussman, J. Geetter, Privacy and big data. Computer (Long. Beach. Calif.) (2014). https://doi.org/10.1109/MC.2014.161
73. H.B. McMahan, E. Moore, D. Ramage, S. Hampson, B.A. y Arcas, Communication-efficient learning of deep networks from decentralized data. J. Am. Stat. Assoc. (2016)
74. L. Xu, C. Jiang, J. Qian, J. Li, Y. Zhao, Y. Ren, Privacy-accuracy trade-off in differentially-private distributed classification: a game theoretical approach. IEEE Trans. Big Data (2017). https://doi.org/10.1109/tbdata.2017.2777968
75. P. Ah-Fat, M. Huth, Optimal accuracy-privacy trade-off for secure computations. IEEE Trans. Inf. Theory (2019). https://doi.org/10.1109/TIT.2018.2886458
76. D.C. Nguyen, M. Ding, P.N. Pathirana, A. Seneviratne, J. Li, H.V. Poor, Federated learning for internet of things: a comprehensive survey. IEEE Commun. Surv. Tutorials (2021). https://doi.org/10.1109/COMST.2021.3075439
77. W.Y.B. Lim et al., Federated learning in mobile edge networks: a comprehensive survey. IEEE Commun. Surv. Tutorials (2020). https://doi.org/10.1109/COMST.2020.2986024

Chapter 2
Internet of Things, Preliminaries and Foundations

This chapter mainly presents a detailed discussion of the IoT technologies and dependent systems with the main objectives of emphasizing the main attributes of IoT systems that might possibly threaten the security of the system. Firstly, the definition and of the IoT system and the detailed description of its architecture are presented along with a taxonomy for dividing its architecture into layers with different complementary roles. Secondly, the concepts of cloud computing, fog computing, and edge computing are discussed and compared in view of IoT systems. Finally, the learned lessons are summarized and pointed out in the last section of this chapter.

2.1 The Architecture of IoT System

The IoT system is typically defined as the interconnection of massive heterogeneous devices and systems in diverse communication patterns including human-to-thing communications, human-to-human communications, or thing-to-thing communications [1]. The structural design of any IoT system is constituted by the integration of physical objects into any kind of communication network and empowered with a variety of computational resources with the objective of developing smart applications and intelligent services for widely distributed organizations or human beings. IoT transforms a physical object from a standard object to a smart object by exploiting a variety of technologies (i.e., artificial intelligence, networking technologies, pervasive computing, Internet protocols, and ubiquitous computing). The design of a reliable and trustworthy IoT system is critical in either industry or academe because of the sensitivity and prominence of dependent applications or services, which are essential

Electronic Supplementary Material The online version of this chapter
(https://doi.org/10.1007/978-3-030-89025-4_2) contains supplementary material, which is
available to authorized users.

to facilitate the realization of smart city notions beyond billions of communicating IoT devices.

Till this time, the absence of standardization and regularity in IoT solutions throughout the public is mostly caused by some matters associated with compatibility, controllability, and interoperability [2]. Similarly, inconsistency in the structural design of IoT systems and the corresponding stacked layers have been demonstrated in the different research literature. For example, according to [3–6] the structure of an IoT system normally entails three primary layers, specifically, the perception layer, the network layer, the application layer (see Fig. 2.1). Besides, the authors of Ref. [7] consider four-layered IoT architecture, where is semantics layer was added to the beforementioned three layers be responsible for managing the business logic of the IoT system (see Fig. 2.2). Other studies [1, 8] deliberate IoT architecture in terms of five building layers including the perception layer, network layer, middleware layer, application layer, and business layer. Thus, it was thought that the non-uniformity and the lack of standardization is the main reason that makes the academic or industrial community still unable to approve a specific IoT reference architecture (see Fig. 2.3). However, the current efforts for determining the layered architecture of IoT systems still providing a coarse-grained definition for system components and their responsibilities.

To mitigate this non-uniformity, the structure of the IoT system is further taxonomized into more fine-grained layers to gain a more straightforward and robust analysis, as displayed in Fig. 2.4. The description of different layers and the constituting components are clearly discussed in the subsequent subsections.

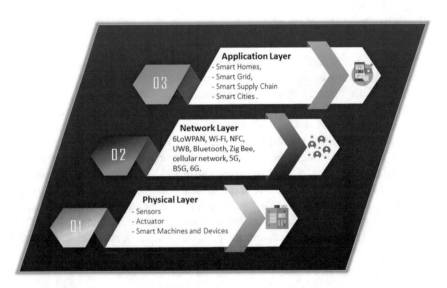

Fig. 2.1 Three layered architectures of IoT system [3, 4]

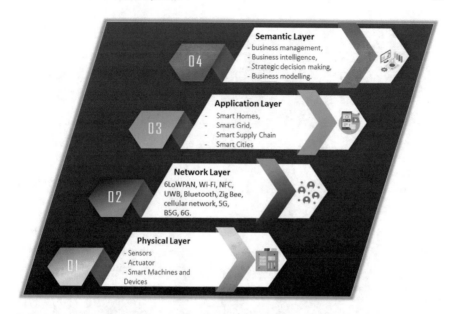

Fig. 2.2 Four layered architectures of IoT system [7]

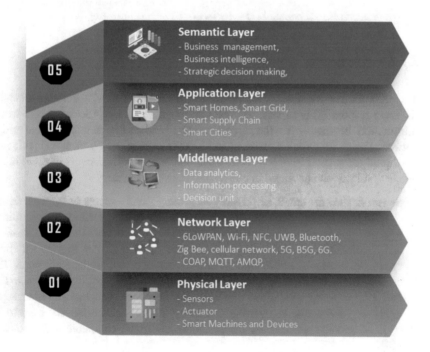

Fig. 2.3 Five layered architectures of IoT system

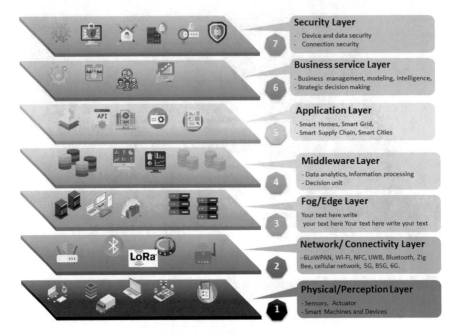

Fig. 2.4 Seven layered architectures of IoT system

2.1.1 Perception/Physical Layer

The perception or physical layer contains the physical IoT sensing components with the main responsibility of sensing, aggregating, and perhaps processing real-world information. In other words, the perception layer behaves like an intermediate bridge between the real and digital environment by transmuting analog measurements into the digital format and contrarywise [9]. The IoT sensing components inserted in the perception layer can be found in one of the following forms:

- **Sensors**: They are small, or possibly incredibly small gadgets or systems developed to identify and measure the differences in a particular kind of streamlined information in the surrounding environment (i.e., temperature, humidity, motion, and acceleration). Usually, the size of these sensors is very tiny and thereby consumes or requires little amount of power to accomplish their missions. The recent advances in sensing technologies improve the capability of sensors in detecting different modalities of environmental or physical factors either separately or jointly, then converting them into digital signals that are easy to be realized and/or processed by the communication and computational resources [10].
- **Actuators**: The actuator is defined as a type of engine or a portion of the smart devices that is accountable for regulating or performing some actions in an IoT system. It takes the data aggregated by the sensors (e.g., pressure, other sources

of power) and transforms it into activities to regulate a system. Data collection from sensors is useless except when this data is turned into intelligence to control or regulate an environment, hence the actuators perform a critical function as elements of IoT systems [11]. The perception/physical layer might contain different types of actuators which are briefly discussed as follows:

- Electrical Actuators: they represent the gadgets driven by slight motors that translate energy to mechanical torque and are utilized to manage various valves per in engines.
- Mechanical Linear Actuators: the act to transform rotary motion to linear motion using conversion devices like screws and chains.
- Hydraulic Actuators: they are defined as minimal devices with mechanical components that are used on linear or quarter-turn valves. It contains a fluid motor or cylinder that employs hydraulic energy to empower a mechanical activity.
- Pneumatic Actuators: they operate in a similar way as with hydraulic actuators however they employ compressed gas rather than liquid.
- Manual Actuators: they make use of levers, gears, or wheels to support motions instead of an automatic external power source.

- **Machine and Devices**: They represent the smart devices that contain the actuators and sensors as embedded components. The IoT systems have no restrictions about the number, position, or distance between multiple devices that can be dispersed throughout the globe [12].

The plug-and-play methods should be operable in this layer to enable configuring a diverse set of sensors. The IoT devices in this layer are typically known to be resource-restricted devices as they come up with small disk or memory size, limited processing capability, short-range battery capacity. However, the ever-increasing number of IoT devices is the main reason behind the continuous massive enlargement in the volume of daily generated data, which indicates a constructive correlation between the IoT systems and the concepts of big data. interpreting the sensory IoT data delivered in the perception layer is a crucial phase in realizing context-conscious IoT applications/services [9, 10, 12]. Thus, efficient analysis of the big IoT data is essential for enhancing the reliability of decision-making at subsequent layers and enable realizing a more secure implementation of IoT systems.

In this context, the deep learning approaches can play a promising role in the authentication of the perception/physical layer. Traditional techniques for physical/perception layer authentication usually apply hypothesis judgments and communicate the arbitrariness and exclusivity of the radio channel between Things to identify deceiving attackers through an IoT network. However, these methods are demonstrated their ineffectiveness, especially under dynamic networks.

2.1.2 Connectivity/Network Layer

One of the foremost aims of any IoT system is realizing the cooperative connection between heterogeneous IoT physical devices to enable the development of smart applications or services at a large scale [13]. Generally, the majority of physical IoT devices are known to be resource-constrained in nature thus, the connectivity/network layer comes to connect them under a lossy and noisy communication setting with low-power resources [13]. However, the deployment of such devices in IoT system often impose multiple connectivity challenges as follow: (1) affording exclusive internet protocols (IPs) to millions of internets connected IoT devices. This could be alleviated by incorporating IPv6 over Low-Power Wireless Personal Area Networks. (2) The design of energy-efficient communication for transmission of data captured by sensing devices. (3) Realizing operative routing protocols that deliberate the limited resources of physical devices and promote the mobility and elasticity of smart devices.

The connectivity/network layer, take on the responsibility of communications between the devices of perception/physical layer other layers of the IoT system. These communications can take place under one of two common internet protocols namely Transmission Control Protocol (TCP) or User Datagram Protocol (UDP). Where the gateways are employed for connecting the Local Area Network (LAN) and Wide Area Network (WAN), hence delivering a route for transmitting the IoT information through a variety of protocols. The connectivity/network layer can integrate a variety of networking technologies such as:

- **Ethernet**: It denotes a networking technology for connecting IoT devices with different parities of wired LANs in IoT systems using physical or hardware tools (i.e., cables, hubs, switches, etc.). Ethernet defines the way the IoT devices can organize and communicate the data with other devices where the data later received and possibly stored or processed [14].
- **IEEE 802.11 Wireless Fidelity (Wi-Fi)**: it is the most common and versatile wireless technology that was developed with the aim of substituting the wired connectivity of Ethernet with wireless LAN communication for covering indoor broadbands using wireless routers and wireless access points. Its goal was to deliver regular, easy to deploy, and use wireless connectivity with cross-vendor interoperability. Wi-Fi is an apparent preference for indoor IoT connectivity owing to the ubiquitous Wi-Fi coverage within buildings (e.g., smart homes, companies, etc.), nevertheless, there is no guarantee to be the best choice for all time [15].
- **Bluetooth**: It is broadly used networking technology to offer inexpensive and short-range wireless communication between adjacent IoT devices (smartphones, personal digital assistants, laptops, etc.) running over an unlicensed band. It has demonstrated itself as a prominent technology for personal area networking (PAN) operating on a little amount of power and concurrently shares a small amount of data [16, 17].
- **Ultra-wide Bandwidth (UWB)**: A short-range networking technology that supports wireless communications across devices through radio waves. Different

from its counterparts, it runs on a very large bandwidth (in the range of GHz) which makes it capable to efficiently capture exactly positional and directional data [18].

- **Radio-frequency Identification (RFID)**: It is a kind of wireless connectivity technology that employs radio waves to inactively distinguish a tagged object. It consists of two elements **namely the** readers and tags. The reader represents the devices with one or many antennas to transmit radio waves and receive signals back from the RFID tag. The tags make use of radio waves to convey the corresponding identity and relevant information to adjoining readers either passively (powered by the reader) or actively (battery-powered) without the need to be in direct line-of-sight with the targeted reader [19].
- **ZigBee**: It represents networking technology that promotes advanced wireless connectivity using low-power digital radio signals to offer limited data transfer rate conforming to the IEEE 802.15.4 wireless standard. ZigBee is developed with a major emphasis on home automation and has demonstrated incredible success for short-range applications [20, 21].
- **Near Field Communications (NFC)**: An emerging contactless networking technology for transmitting the data between different NFC compatible devices through electromagnetic radio fields. It enables the users to just wave the smart device over another IoT device to send information with no touching or connection setting up requirements. Unlike RFID, NFC operates only on very short distances (i.e., inches), whereas NFC devices can operate either as a reader or as a tag [22–24].
- **LPWAN (Low Power Wide Area Network)**: A networking technology designed to satisfy the IoT requirement over long distances. The LPWAN devices can remain operating for a long time (years) while maintaining a slight energy consumption. Nevertheless, it can transmit signals to deliver exact information between different IoT parities over a lengthy intervallic period [25–27].
- **Cellular Network**: It represents **a** radio wireless network geographically distributed across multiple regions named cells, where one or more fixed-position transceivers (called base stations) are employed to serve each cell. These cells collectively afford radio coverage over wider geographic zones, meanwhile, the frequency of a particular cell varies from those of adjacent cells, to assure the quality of service per each cell and evade intervention issues. Hence, IoT devices are able to communicate even they are moving between cells throughout transmission [28, 29].

The Connectivity/Network layer also accommodates a variety of messaging protocols to enable seamless data sharing. The most popular protocols for IoT system are pointed out as follow:

- **Data Distribution Service (DDS)**: It represents is connectivity protocol for data-centric communication from the Object Management Group® (OMG®). It unifies various system components delivering scalable design, reliable, interoperable, and low-latency data exchange to satisfy the requirements of mission-crucial IoT applications or businesses [30].

- **Message Queue Telemetry Transport (MQTT)**: It represents standard messaging protocol for IoT systems as presented by the Organization for the Advancement of Structured Information Standards (OASIS). It was devised as an exceptionally lightweight and efficient publish/subscribe messaging exchanging protocol that is perfect for linking distant devices with small resources and minimum network bandwidth. Currently, the bi-directional communication ability of MQTT and scalability to link up millions of IoT devices make it a dominant protocol in the majority of IoT applications [31, 32].
- **Advanced Message Queuing Protocol (AMQP)**: It provides an open standard for interoperable exchanging of data messages between different organizations and applications irrespective of the employed platform or message broker vendor. It promotes an easier, more secure way for passing real-time data streams and business transactions among different parties across distributed IoT environments, and also inside mobile infrastructures [33, 34].
- **Constrained Application Protocol (CoAP)**: It defines web transmission protocols specialized for use with resource-limited nodes (e.g., low-power, low memory) and confined networks (i.e., low bandwidth network, 6LoWPANs) in IoT environments. It is specifically designed for machine-to-machine (M2M) applications (e.g., smart energy and home automation) [35, 36].

2.1.3 Fog or Edge Layer

In the initial phases of designing IoT systems, the stationary or mobile IoT device directly contact with a central server to handle their requests, which progressively shown to result in high latency issue. With the continuous increase in the number of IoT devices, latency became a major hurdle of IoT systems. The emergence of fog computing and edge computing concepts (discussed in later sections) presented a distinctive solution for this issue while improving the scalability of IoT Systems. The notion behind fog or edge layer is to provide a decentralized form of computing that brings the computing and storage resources closer to the end-users device thereby enable an intermediate level of computation between centralized servers and network terminals. Hence, this layer permit analyzing and manipulating large amounts of real-time data in the vicinity. Fog or edge layer has come to be the standard for the 5th Generation (5G) wireless networks enabling the IoT systems to connect with extra devices with smaller service latency compared to the dominant 4G standards. Wide-span of operations of IoT networks is taken place at this layer hence saving time, resources and accordingly lead to immediate responses and enhanced performance.

2.1.4 Middleware Layer

A middleware seeks to effectively characterize the complications of IoT systems along with their physical component, hence letting the developers concentrate on solving the problems that did not result in any system failures [37]. Bearing in mind, the majority of those complications are generally reasoned by or associated with computational and/or communication matters. To this end, the middleware layer takes the responsibility of offers a software level amongst applications, the operating system, and the network communication levels; it enables cooperative processing.

According to the computational point of view, the middleware layer act as an intermediate stage between the system software and an IoT application. Its principal responsibilities are briefly debated as follows.

- First, it enables collaboration between heterogeneous IoT elements (either physical or software) in such a way that permit interaction among different IoT parities in an easy and effortless manner, which demonstrates the roles of the middleware layer in realizing the interoperability between the IoT devices.
- Second, a middleware has to offer scalability amongst numerous IoT devices that are expected to co-operate in the IoT system. The upcoming evolution of IoT devices ought to be tackled in the middleware by supporting vital adjustments to cope with the organizations' scalability requirements [25].
- Third, the context realization and device detection, which should be provided by a middleware layer to identify and monitor IoT elements throughout the environments. In other words, the middleware layer is accountable for supporting context-aware computing to interpret and comprehend sensory IoT data. The sensory data generated at the physical layer can be exploited to acquire the context information that followingly can be hired to deliver or allocate IoT services to the requesting clients.
- Finally, the security and privacy of IoT devices can be preserved at this layer as the aggregated data typically belong to either humans or the environment. Hence, intelligent deep learning approaches can be deployed at this level to comply with the privacy and security requirements of the IoT system [9].

2.1.5 Applications Layer

The great advancements in the physical IoT industry and communication technology have been transforming the concept of the internet from a universal system of connected computers to a universal system of connected things. The convergence of CPS and IoT enables intelligent applications that automate tasks that were mainly requiring human interventions in previous times. The valuable insights extracted from the large volume of data and intelligent big data analytics have been paving the way toward the development of a wide variety of IoT applications (i.e., smart homes,

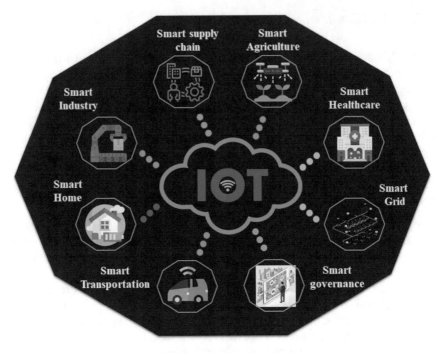

Fig. 2.5 Popular smart applications build on the IoT system

smart transportation, smart healthcare, and smart grid). The following subsections briefly discuss the most popular applications (see Fig. 2.5).

2.1.5.1   Smart Homes

With the proliferation of affordable IoT devices, they become an essential component embedded in smart home appliances such as smart televisions, fridges, air conditioners, heating systems, etc. This enables these appliances to be more aware of the surrounding environments, which offer more intelligent performance in terms of power consumption and easy remote control without requiring much human effort. The integration of the smart and connective appliances facilitates the design of a fully automated smart home system to interpret and act in response to surrounding actions and states. For example, opening or closing the door according to the recognized face, turning the air conditioners on or off according to weather states, controlling the television or lights with human gestures, detecting gas leakage, and baby monitoring. Smart homes' system ought to constantly take into consideration the indoor and outdoor environmental conditions [38], where the indoor conditions involve the states of all the internally controlled appliances (i.e., television, inner doors, smart lights, etc.), and the outdoor conditions represent the exterior things or states that are not under the control of the smart home system, but they have an essential role in the design of such system (i.e., electricity, weather, etc.). In the nutshell, the emergence of IoT system have been demonstrated great success in improving the quality of human lives in smart home and pave the achieve this success in a wide variety of indoor environments (e.g., smart buildings, companies, business offices, agencies, etc.). However, the privacy and security concerns of users still the main challenges facing the design of any smart indoor system.

2.1.5.2 Smart Healthcare Applications

Recently, a growing desire has been witnessed to make the healthcare environments (e.g., hospitals, medical institutions, health centers) more tailored, cost-efficient, and dynamic which coincides with advancement in IoT technology making it the most promising solution for satisfying the emerging requirements of healthcare systems and thereby improving the quality of medical services. The introduction of IoT system into medical or healthcare environments bring us with a new subset of generic IoT platform called the Internet of Medical Things (IoMT) [39], or sometimes referred to as the Internet of Healthcare Things (IoHT) [40]. Different from the generic IoT system that mainly depends on Wireless Sensor Networks (WSNs), the IoMT mainly relies on the Body Sensor Networks (BSNs) which is a network established using a set of connected sensing devices deployable at different parts of the human body. Accordingly, the patients linked to such network can be easily monitored or traced for any variation in the corresponding health conditions (i.e., biometric evidence, vital signs), which typically guarantee to improve the quality of diagnostic procedures and medical services [41]. The IoMT devices are deployed in healthcare environments to carefully monitor and recording patient information and transmit alarms to the responsible healthcare applications in crucial situations to enable delivering fast and timely medication for patients. IoMT devices have been widely spreading across a huge portion of the medical workspace and have been shown to have a considerable role in revolutionizing the medical/healthcare environment from unorganized and manual healthcare systems to coordinated and automated systems.

Nevertheless, compared to the other kind of applications, the IoT-based healthcare systems/ applications have strict security requirements as the patients' data and medical transactions are very sensitive to a degree that might threaten the safety of patients' health especially under emergency situations. For example, when the patients with attached IoMT device have undergone an emergency case, they should be acknowledged into a hospital as rapidly as possible, while stable cases can be served at the regular visits. According to this, the smart healthcare system can easily take care of all patients by accessing the implanted IoMT devices. Moreover, it is noticeable that the absence of cybersecurity at various sections of IoHT or IoMT applications is demonstrated as multi-faceted security defects with implications fluctuating from privacy leakages to jeopardizing patients' lives. Therefore, complicated security constraints become unsatisfactory and impractical, which necessitate an intelligent security approach that carefully takes into account the tradeoff between the security and latency of making decisions throughout emergency circumstances [42].

2.1.5.3 Smart Transportation Applications

The contemporary innovations in vehicular technology have been greatly improved the methods of building, operating, and controlling today's vehicles to be ubiquitously computerized and network-connected [13]. The pervasive and connected nature of smart vehicles makes them easy to integrate into IoT platforms as they are usually armed with many embedded units like engine control module (ECM), wireless-enabled onboard units (OBUs), electronic controlled units (ECUs), application units (AUs), trusted platform modules (TPMs), head units (HUs), Closed-circuit television (CCTV), a global positioning system (GPS), magnetometers, accelerometers, and gyroscope. Each unit is responsible for collecting a particular kind of vehicular information while performing some tasks within a connected vehicle [43]. For example, The ECUs usually aggregate the data about vehicular dynamics (e.g., velocity and locations information) by swapping messages with an AU and OBU across the vehicular network, the AUs are accountable for implementing applications delivered by a variety of remote service providers (RSPs), while the TPM often take the responsibility of storing certificates, and cryptanalytic keys for securing the communications. Besides, the smart vehicles are supplied with a diversity of wireless networking technologies that empower vehicles to communicate with the neighboring vehicular agent such as roadside units (RSUs), smart traffic lights. Considering the convergence of IoT and vehicular technology, the smart vehicles own powerful computing facilities that qualify them to be perfect nominees for on-road smart transportation applications, such as traffic prediction, autonomous driving, localization, and tracking.

The emergence of smart transportations and autonomous vehicles impose many considerable constraints in terms of traffic security and system malfunction while providing various advantages. These services will have a substantial influence on the citizens' lives and society because of the total reduction in the ratio of traffic crashes, finetune the road planning, optimize traffic flow, conserving the sustainability of the environment, and satisfy the different necessities of drivers [44]. According to this, it worth observing that the applications and services of smart transportations indispensable steps toward the realization of smart cities.

2.1.5.4 Smart Grid Applications

A prolific growth has been witnessed for the global energy demand, and electric power is estimated to constitute up to 40% of the overall power production to satisfy the universal expanding power utilization needs by 2040. Achieving the dependability in the standard grid system becomes more challenging because of several reasons such as grid bottleneck, bigger energy transmissions over remote distances, mature infrastructure and unsatisfactory venture on maintenance, cumulative electrical power utilization at the time of peak power request, and amplified consumption of the distributed resources. The deficiency in electric energy has a catastrophic influence on industries, markets, organizations, and all countries' sectors. The traditional power network becomes inefficient because of reliance on conventional methods of processing and maintenance [45]. Smart Grid (SG) development has continually had a great emphasis in both industry and academia because of the growing complications of electrical energy systems, daily increasing needs of electricity, and the obligations of abundantly dependable, effective, and guaranteed power support. Accordingly, the smart grid becomes the backbone for the subsequent generations of power systems where the electricity and consumption information is transformed in a bi-directional way [46]. The integration of IoT systems empowers the capability of smart grids for system management, data amalgamation, dependable data messaging, and improved analytical services, which in turn facilitate handling the transaction of power suppliers and satisfying the consumer demands in such a way that optimize both energy utilization and expenses. The application of IoT platforms in the smart grid improves the monitoring capabilities of the absolute power network from the consumers to the energy suppliers using the diversity of applications and services [2]. Scrutinizing the load feature and power expenditure at the consumer end is a challenging task for the traditional grid, the smart grid enables handling this concern using the bidirectional communication between the service providers and customers. Furthermore, the capability of communicating with consumers' IoT devices enables adaptive modifications based on on-peak or off-peak times. Thus, the IoT enabled smart grid applications to have the potency in supplying dependable power resources.

2.1.5.5 Smart Manufacturing Applications

The requirements for sustainable, intelligent, personalized, and products lead to the emerging of new smart manufacturing paradigms Industry 4.0 schemes, were machines with a particular level of connectivity ability that enable them to collaborate for exchanging massive manufacturing data over industrial networks. The deployment of IoT in the industrial environment brings the academic and industrial community with a specialized variant of IoT system known as the Industrial IoT (IIoT), where the connected manufacturing machines and industrial gadgets are the main "Things" constituting up the system [47]. A wide variety of smart industrial applications can be easily developed over IIoT systems to effectuate the interoperability of industrial data and systems for elastic and optimized allocation of resources, automatic accomplishment of procedures, coherent optimization of processes, and fast adaptation of environments. The IIoT abstracts manufacturing procedures into data utility, transformed industrial agents into data sources and local decision-makers, accumulates the elementary manufacturing data in all directions, and syndicates the prevailing computational resources to make in-depth data mining and analysis to optimize the manufacturing processes [48]. Thus, the smart industry applications introduce over-whelming improvements in the production, business, and executive strategies of the industry, leading to a robust basis for the reasonable distribution of resources and enhancements in quality of production and supported service. However, the current and evolving strategies of Industry 4.0 manufacturing needs efficient mechanisms to satisfy the complex security requirements. The majority of the present smart industry applications generally prone to the large surface of security vulnerabilities where the manufacturing processes and/or data are subject to be attacked or hampered by intruders. Flawed and damaged data is likely to cause inaccurate decisions and present a considerable threat to interconnected complicated manufacturing systems [49].

2.1.5.6   Smart Agriculture Applications

Motivate by the great benefits gained by IoT systems in the industrial field, the appli-cation of IoT systems extended to cover the agriculture environment. The implanted connected IoT sensors across different agricultural provide take the responsibility of collaborative sensing and monitoring different aspects (i.e., weather states, humidity degree, temperature degree, and moisture degree, etc.) of the agriculture environment [50]. The accumulated data is subsequently analyzed to gain insights critical for the development of smart agricultural applications and services including automated irrigation, examining the quality of water, monitoring soil ingredients, infection, and insects, and detecting crop diseases [51].

2.1.5.7   Smart Supply Chain Applications

The definition of a supply chain has been evolved from just the flow of mate-rials between customers and providers to involve the flow of finances, services, and information through a network of connected entities. The supply chain process is a complex and challenging task that necessitates active communication and a sympathetic organizational background, even though it contemplates a single site plant supplied with a limited number of suppliers. The supply chain represents the foundation of the present consumer first globe with the main aim to capitalize on customer benefit and achieve a competitive lead in the marketplace [52].

With the introduction of IoT and communication technologies, the efficiency and effectiveness of supply chain applications have witnessed great improvements and scale up to cover manifold geographical frontiers in various socio-economic aspects, each forcing its certain set of examinations and equilibriums to guarantee easy opera-tion for the chain. Accordingly, the supply chain has become a prevailing and essential requirement for a wide variety of fields, such as the agricultural industry, the health-care industry, the vehicular industry, and the smart grid industry, etc. Thus, it is vitally important to maintain the supply chain operating without interruptions via different security tests. Nevertheless, the ever-increasing variables in the prevailing system have complicated security operations. The excellence of products degrades considerably prior to reaching the customers because of some inadequacies in the existing supply chain system [53]. In Addition, the common security flaws of the supply chain system have supplemented the problems of forgery and pirated commodities, which in turn result in a great financial loss on the consumer's side and lead to a bad reputation for the companies or provider. The complex characteristics of the supply chain make it difficult to follow and go back over every single stage per the chain. Numerous perpetrators exploit such weakness and initiate in variety of attacks and malware to malevolent and spiteful actions.

2.1.5.8 Smart Governance Application

Like the previous applications, the integration of IoT system into governmental operations offer great flexibility and synchronization between different multi-disciplinary governmental sectors. The introduction of smart governance applications is inevitable to realize smart cities. Collaborative exchange of data between governmental sectors enables the authorities (i.e., ministries) to obtain plentiful information from the versatility of sensory data that enables mitigating the shortcomings of traditional governmental management systems in an extraordinary style [54]. The aggregated information can be easily fed into a knowledge-based system for intelligent mining, compilations, analysis of correlated and autonomous data at various sectors, which finally offer multi-perspective insights that help the authorities finding the optimal decisions and predicting the possible consequences. To sum up, smart governance paves the way toward more intelligent political management, police system, military services, disaster management, etc. However, the privacy and security of governmental or political data are of extreme importance as it usually contains very special and critical information [55].

2.1.6 Business Service Layer

The business layer takes the responsibilities of handling the service, functionalities of the complete IoT system, including business management, applications, revenue models, and customers' private information. In particular, the information generated from the aforementioned layers gets analyzed at different levels to gain deep insights that become valuable only when it provides a solution for a certain problem or facilitates realizing some business objectives. Hence, the stakeholders should get involved and collaborate in making decisions to benefit from the generated insights to improve the system's productivity.

The process of decision-making typically entails multiple individuals laboring with multiple and possibly different software tools, which motivate separating the operations of the business layer from the operations of the application layer. The conducted petabyte-scale analytics at this layer is dominated by the conformity and record withholding procedures. Artificial intelligence algorithms are typically deployed on this layer to improve business development, functioning optimization, drilling insights. Furthermore, metadata and reference data executive, business rules administration, and the operative health of the previous layer also considered through this layer. Finally, the business layer serves the end-user by simplifying the design of systems that support the provision of business solutions regardless of technical limitations.

2.1.7 Security Layer

The rapid proliferation of IoT technologies and the increase in the range of coverages of IoT networks indispensably enlarge the attacks surface and make the IoT system more vulnerable to different threats (i.e., Data breach, malware, anomalies, etc.) that might cause catastrophic consequences. Accordingly, security and privacy become vitally important elements of any IoT system. Thus, the security is integrated as the final layer of the architecture of the IoT system that is responsible for satisfying the security requirements at different layers, detecting, and defending the IoT system against possible cyber or physical threats without affecting the quality of IoT applications or supported service [47]. In view of this, different categories of security are introduced to protect the IoT system at different levels.

- Device security. Recent vendors of IoT devices normally apply security characteristics for the hardware and firmware they contain. This incorporates entrenched Trusted Platform Module (TPM) chips with cryptographical signature for the purpose of authenticating and protecting the IoT devices. In addition, a reliable boot procedure inhibits the run of any unauthorized code on an operated-up device. Other methods regularly update the security patches for protection. metallic guards are often applied for physical protection to prevent physical entry to the device.
- Connection security. Since the sensed or captured data are typically broadcasted over networks (wired or wireless), the communication or networking security must be assured to prevent access to sensitive information during transmission. Advanced IoT message exchanging protocols (e.g., DDS, MQTT, AMQ P, etc.), can use standard Transport Layer Security (TSL) protocol to perform Encryption, Authentication, and Integrity check to guarantee end-to-end security of data during communications over the Internet.
- Cloud security. The final location for data storage ought to be also encrypted to alleviate threats of revealing confidential information to invaders. Securing cloud servers can entail authentication and approval procedures to restrict access to the IoT service and applications. One more essential security technique is the control of the identity of the device to confirm its reliability before permitting it to access the cloud [39].

Fortunately, the current IoT solutions supplied from international providers, (e.g., AWS, Cisco, Microsoft, etc.) come with pre-developed security procedures involving access control, device authentication, and data encryption. Nevertheless, the rapid evolution of attacking techniques and patterns makes the standard security mechanism insufficient and imperfect to cope with this rapid evolution. This motivates academia and industry to move toward exploiting the machine learning algorithms to develop intelligent security solutions by learning different attack patterns from IoT traffic data.

2.2 The Cloud Computing Based IoT System

Cloud Computing can be defined as a service model where user devices are able to connect to cloud servers. The servers come up with a considerably extensive pool of computational resources, storage, communicational resources, and energy resources which are extremely powerful than the user-side resources. Hence, the remote cloud servers are able to easily perform big-data analytics, decision making, and computations necessary to satisfy the customer's requirements asks that go beyond the capabilities of local devices and deliver these tasks as on-demand applications/services for the customers throughout IoT networks. According to the research and markets report [56], the cloud computing market is poised to grow by $287.03 bn during 2021–2025, decelerating at a Compound Annual Growth Rate (CAGR) of over 17% throughout the prediction time, where multiple companies have been recognized as key cloud service providers, such as Adobe Inc, SAP SE, Salesforce.com Inc, Microsoft Corp, International Business Machines Corp, Hewlett Packard Enterprise Development LP, Oracle Corp, Amazon.com Inc, Alphabet Inc, and Alibaba Group Holding Ltd [52].

Standard Cloud Computing often denotes the situation where the customer devices are stationary computers (desktop, PC) located in a specific location, while Mobile Cloud Computing indicates the case where the customers' have mobile devices to access the internet. The introduction of mobile devices makes the service conditions substantially changed. Stationary computers occasionally could utilize wired connections to communicate with the cloud servers (small distance scenarios), whereas mobile devices every so often uses cellular or wireless networks to communicate with each other or with cloud servers over the Internet. In Addition, the resources of mobile devices are usually more constrained than static computing devices, thereby have little chance to locally execute a small set of limited resource applications [57]. This indicates that mobile devices are likely to benefit more from cloud resources to remotely run complex applications. However, the requirement for powerful wireless communication and the limitations of local devices obscures the deployment and the provision of cloud services. Additional obstacle, variety of mobile applications (i.e., Virtual Reality, Augmented Reality) necessitate real-time response which cannot be conceivably provided by Mobile Cloud Computing because of long distances between the cloud servers and customers' devices that might be situated in dissimilar countries or continents [58].

Currently, the IoT system has been considered as a major constituent of the forthcoming Internet that attracted growing interest from either academics or industries because of the relevant robust capabilities for providing thrilling smart applications over robust communication technologies. The IoT system impeccably intercommunicates heterogeneous things (IoT devices) to create a physical environment for automatic sensing, storing, manipulating, and exchanging information related to the targeted applications with little or possibly no human intervention. Nonetheless, the continuous increase in the number of IoT devices leads to exponential growth in quantities of data leading to more complexity in assuring the Quality of Service (QoS) requirements because of resource limitations of IoT devices. In the Meantime, cloud computing possesses the powerful resource capability necessary to deliver on-call, robust, and efficient services for IoT use domains. In particular, the convergence of IoT systems and cloud computing blaze a trail toward a new concept called a cloud of things (CoT) [57].

Indeed, the abundance of pooled resources accessible from the cloud servers is extremely valuable to IoT systems, while the cloud can obtain further acceptance in real-world applications from incorporating it into IoT systems [43]. Accordingly, the CoT can transform existing delivery models of IoT services with negligible executive exertion, high efficiency, and service accessibility, and convenience. Figure 2.6 typically illustrates the design of a CoT system consisting of cloud servers, IoT devices, application layer, and analytic services. In this hierarchy, IoT devices are employed to sense and accumulate the environmental measurements then upload the data to the cloud servers for acquisition purposes because of their limited storage and computations. Cloud servers take on the responsibility of big data storage and subsequent processing, and analysis. Analytic services can be provided to support IoT systems, such as statistical analysis, historical records, information storage. The cloud processes the collected data based on the necessities of different smart applications and offers them results as IoT services that can be easily delivered to the end-users at high quality.

Generally, instantaneous services become more realizable anywhere and anytime with the emergence of the CoT system thanks to automated resource delivery capacities which facilitate the provision of self-directed service without the need for human engagement. Cloud computing promotes unrestrained virtualization facilities which further paved the way toward improving IoT resource management by remotely offload the data and processing in an elegant manner. This not only improves the computation abilities of local devices but also addresses effectively issues of IoT systems in terms of energy-saving and bandwidth preservation. Significantly, the CoT system enables streamlined and automated IT conservation and executive solutions by exploiting, virtualization, resource infrastructure, and cloud servers. IoT users can communicate effortlessly with cloud computing to perform their functionalities without any necessity for hardware/software deployment or human engagement. Furthermore, cloud computing also makes the system management facilities available to provides glowing and boundless connectivity and communications among different IoT devices, smart things, and customers to support pervasive applications, which would endorse the inclusive partnerships of manifold IoT systems in

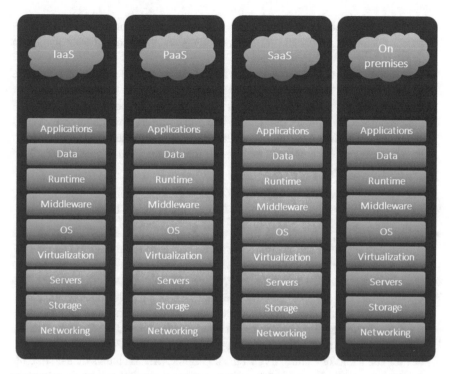

Fig. 2.6 Comparison between SaaS, PaaS, and IaaS, especially in whatever is presented by the users and what is provide by the Service Providers per every service model

the imminent Internet. The data-hungry nature of deep learning development makes cloud computing is the best choice to design and train a complex deep learning model using larger datasets in a considerably short time.

Commonly, cloud computing can present in one of three fundamental service models as followingly described (see Fig. 2.6).

- Infrastructure as a Service (IaaS): A service model that involves the provision of all resources including operating systems, network, computing, and storage throughout internet connectivity as components of an on-order service according to the pay-as-you-go rule. This enables the customers/end-users to escape purchasing servers, storage, or software and as an alternative acquire, scale, and shrink these resources in accordance with the as-required basis.
- Platform as a Service (Paas): A service model that eliminates the necessity for organizations to operate and control the software infrastructure (hardware and operating systems) by managing the application infrastructure on the cloud and enable the and customers to emphasize the deployment and management of their applications. In other words, it delivers a platform for software development as a service over the Internet including middleware, development tools, database, and middleware. This enables being more efficient it alleviates the necessity to

be concerned about capacity planning, patching, software maintenance, resource procurement, or any of the other indistinguishable weighty tasks engaged in operating customers' applications.

- Software as Service (SaaS): A service model involves the licensure of a software application to customers. Licenses are normally supplied either on-demand or according to the pay-as-you-go rule. It is responsible for providing clients with a finalized software product that is operated and controlled by the provider which alleviates the burden of service maintenance or infrastructure management from the shoulders of the end-users and makes them focus on the specific chunk of software it concerns about.

The CoT system could be under any of the four deployment paradigms namely Private Cloud, Pubic Cloud, Community Cloud, and Hybrid Cloud.

- Public Cloud Deployment: As the name indicates, public cloud deployment is extremely suggested and is simply comprehensible to the public. It is a picture-perfect choice for corporations with expanding and unpredictable requirements and also recommended option for enterprises with minimal security worries. The provided services are maintained and regulated by cloud providers, organizations, and some grouping of business leaders. Thus, the customers pay a cloud service provider to procure the networking services, virtual computing & storage accessible over the internet. This offers a huge supply model for the squads working in development and/or testing. Its design and implementation are swift and straightforward, rendering it a perfect preference for test ecosystems. The main benefits gained from public cloud deployment can be briefly discussed as below:

  - Minimal Investment: The customers pay only for the utilized resource, scaling up and down according to the customer needs. It promotes neither procurement of physical hardware nor waste of computation with the exception of the hardware to communicate with the cloud.
  - Elasticity: The customers can act in response to traffic thorns immediately. The customers can also design software solutions to enthusiastically expand or shrink resources to cope with peak loads.
  - Core Competency: The data center and infrastructure management is a key benefit of the cloud.

On the other hand, the main drawbacks of the Public Cloud are the security and privacy of data because of the lack of protection from cyber-attacks leading to a wide range of vulnerabilities. Another possible limitation may be the communication overhead under an excessive number of requests.

- Private Cloud Deployment: There is a bit to no discrepancy between a public and a private deployment according to the technical perspective because their architectures are almost identical. Nevertheless, as contradicted to a public cloud that is accessible for the public, only one particular corporation possesses a private cloud. The server can be hosted outside or on the premises of the owner corporation. Irrespective of their physical site, these infrastructures are retained and managed over a defined private network and make use of the hardware and software that are planned for usage only by the organization. contrasted to the public deployment, the private cloud offers broader chances for modifying the infrastructure according to the corporation's needs. A private model is remarkably appropriate for companies that aim to protect their mission-crucial functions or for businesses with continuously altering conditions [4].

- Community Cloud Deployment: A community paradigm mostly seems like a private deployment but differs only on the group of involved customers. While a single organization possesses and manages the private cloud, and the other organizations with analogous conditions share the infrastructure and associated resources of a community cloud. When overall the contributing organizations have standardized security, privacy, and execution obligations, this multi-resident data-center design supports the organization to improve their productivity.

- Hybrid Cloud: As is typically the situation with any hybrid experience, a hybrid deployment incorporates the finest features of the aforementioned deployment paradigms. It permits the organizations to combine and match the aspects of the three paradigms that efficiently cope with their needs. For instance, an enterprise can balance its burden by placing mission-crucial workloads on a reliable private cloud and deploy fewer private tasks on the public cloud. The hybrid cloud deployment model not only protects and regulates tactically valuable assets but also performs in a cost-efficient and resource-useful manner while facilitating the portability of data and application.

2.3 The Fog Computing Based IoT System

The paradigm of fog computing was originally presented at Cisco Inc by 2012 [59]. The IEEE standard 1934–2018 [60] defined fog computing as "a system-level horizontal architecture that distributes resources and services of computing, storage, control, and networking anywhere along the continuum from cloud to Thing. In this context, fog computing considered as a modern computing model that transfers responsibilities from the cloud servers/datacenters to a lighter set of servers geographically close to network edges that collectively constitute a fog layer to be deployed, which consists of some servers deployed between the cloud backend and local IoT devices [47]. This means that fog computing involves pushing intelligence down to the LAN level as well as process the local data at IoT gateways. The major objective behind the design of the fog computing paradigm was to expand cloud-provided services and functionalities including computations, storage, database transaction, amalgamation, security, and managing IoT devices, exploiting the vicinity of the fog layer from the edge of the network. In this way, fog computing is demonstrated as a successful and promising solution for improving the quality of service in IoT environments because of the thrilling advantages of lessening network bottleneck, end-to-end latency, communication overhead, enlightening security and privacy, and improving the system scalability.

Moreover, there are allegations within the production of the massive industry prospects that could be originated with the introduction of fog computing in IoT. With the efficient allocation of networking, processing, storage, and controlling service alongside the Cloud-to-Things field, fog computing can converge the obligations of current applications for resource pooling, local substance, and instantaneous computing. Intrinsically, Fog computing has been attracting increasing interest in academia and industry. It is a reality that fog computing does not reinstate cloud computing, instead, it completes by offloading data or service demand that can be handled in the vicinity [45]. Nevertheless, the limitations of cloud computing are recognizing and the shortcomings of the cloud computing-based IoT system stress the necessity for integrating fog computing to permit or worldwide pertinency. In the same way, dew computing (DC) is an evolving paradigm that is expected to revolutionize IoT systems. In Ref. [61], DC is supposed to hinge on a micro-service method within an extremely heterogeneous, upright, and distributed order. It paves the way for a unified virtualization-free computational perspective wherever data dispersion into constrained devices is feasible. Therefore, granting data convenience even without constant Internet contact. The excessive scalability and self-managed features of DC make it major for to successful IoT system. Nevertheless, numerous research problems will evolve because of its excessive complication. With fog computing midway role of installing prevailing computing infrastructure in connecting things to the cloud, fog computing will be main to the achievement of current and evolving technologies such as vehicular networks, e smart grids, smart homes, smart cities.

To sum up, the main advantages of fog computing can be briefly pointed as follow:

- Geographical distribution: contrary to cloud computing, fog computing promotes a decentralized and distributed management of resources over d at the edge of networks.
- Heterogeneity, fog computing enables the integration and dealing with a wide variety of computing nodes either static (RSUs, access points (APs), high-end servers, edge routers) or mobile (i.e., smart vehicles, RSU, smartwatches, etc.). It also provides a high degree of interoperability.
- Cloud cooperative, fog computing can operate cooperatively with the cloud to deliver multi-layer and in-depth analytics, where not only delay-serious applications can be deployed at the fog layer, yet computation exhaustive and delay-lenient applications are also applicable at the cloud layer thanks to its rich computational resources. Therefore, fog computing is usually a complement to the cloud instead of an alternative [43].
- Scalability, bringing the computation closer to the edges in the fog computing model facilitates the scaling up of the IoT system by adding extra end nodes.
- The on-premises placement of fog nodes makes them accessible by devices even if the connection to the global network is declining. This could possibly help to investigate occasional connectivity problems and help to secure IoT infrastructure by intelligent security solutions to detect attacks or anomalies.

Fog computing is not a remedy for all the gaps of the CoT and still in the initial stages with limited deployment in the real world scenario. Multiple open challenges still facing fog computing solution in IoT environments that can briefly be described as follow:

- Firstly, Fog-based IoT systems include an additional amount of work for maintenance, and mostly necessitate exceptional training, which implies additional costs compared to the Cloud. While the cloud platform is retained by a dedicated team belonging to the cloud service provider, the Fog computing moves this accountability to customers of the IoT environments. The proliferation of fog computing throughout the IoT system possibly adds more complexity to this task. When a security problem is discovered within a specific slice of fog computing software, the maintainers take on the responsibility of updating their software at all points of the IoT system, rather than having a single update managed by a single entity at the Cloud.
- Secondly, an incident response might be hindered by the dispersed nature of fog computing. This could reveal itself in various methods: essential security knowledge may possibly not be accessible on-site, and expert incident response crews should be asked in from outside companies. More, complicated incidents may possibly necessitate collaboration between various entities with digital forensic evaluation, which could not permanently be feasible or result in a large overhead.
- After All, the compatibility among fog nodes might possibly lead to an enormous issue. When the specifications are poorly defined or not carefully followed, it will be extremely tough to encounter the strict constraints established by IoT environments with nodes from diverse vendors that are unable to interoperate effectively and perfectly. This, in turn, can destructively influence the capability to combine and offload responsibilities to other nodes in the LAN, and possibly contravene security strategies when some nodes are incapable to sustain the essential obligations.

2.4 The Edge Computing Based IoT System

Edge computing was firstly introduced by the European Telecommunications Standards Institute (ETSI). The fundamental concept behind edge computing is to transfer some demands and responsibilities from the core network to the access network, reaching the effective consumption of storage, computing, and communication capabilities. This thought profoundly combines the conventional cellular network with the Internet facility, seeking to lessen the end-to-end latency of provision of mobile service and discover the intrinsic abilities of the wireless network, so improving the customer experience, introducing new variations in the operation styles in the IoT ecosystem [48]. By 2017, the ETSI broadened the model of edge computing from telecommunication cellular networks to other access networks, (i.e., Wi-Fi and fixed access), and became known as multi-access edge computing (MEC).

Edge computing describes a modern computing paradigm for analyzing, processing a fraction of IoT data-consuming the computational and network resources distributed over the edge nodes between physical/perception devices and centralized cloud datacenter. It takes the advantages of limited resources of edge devices to execute local manipulation, make some initial decisions, and then transmit the local data results of these slight transactions to powerful cloud servers. The main advantages of edge computing applied to the IoT system are pointed as follows:

- Enhance the System Productivity: beyond the data collection and uploading to the centralized cloud platform, edge computing facilitates real-time data processing as it effectively lessens the global system delay as well as the need for communication bandwidth.
- Privacy and Security Preservation: Cloud platform service providers give customers a comprehensive system of centralized data security protection solutions. Nevertheless, the leakage of centrally stored data typically results in disastrous effects. Edge computing-based IoT systems help organizations to install extremely suitable security solutions in the local proximity, decreasing the possibility of attacks or data leakage throughout the communication of data to the cloud backend.
- Cost-efficiency: The direct transmission of IoT data to the cloud, data mobility, excellent bandwidth, and IoT latency features necessitate plenty of functional expenditures. Edge computing decrease data uploading capacity, in that way lessening data migration amount, bandwidth utilization, and latency, which imply saving much of costs.

Table 2.1 compares cloud computing, fog computing, and edge computing based on different attributes including Network Access, Mobility Support, Scalability of Servers, Computing Power Location, Reliability, Deployment cost, Maintenance, Standardization, Operation Mode, Service Coverage, Service Coverage, Virtualization, and Constitutive Elements.

2.5 Summary and Learnt Lessons

The most important donation of this chapter is an instructive discussion of IoT systems by defining the structure of IoT systems, indicating its attributes and components that are prone to security threats. Furthermore, beyond the previous discussions, it is concludable that the design and characteristics of IoT systems can raise security hazards due to the following rationalizations.

- IoT systems span a lot of communicating devices that can be constructed to independently acclimate to the neighboring environment. These IoT devices are prone to be operated or managed by other devices []. Thus, intelligent solutions should consider designing end-to-end security rather than maintaining the security of IoT devices autonomously.
- Inherently, the IoT is considered a multifaceted system that entails a variety of components and applications, each one of which has a distinct resource requirement. This nature reveals the vast complication of IoT systems under broad of IoT

Table 2.1 Comparison between different characteristics of cloud computing, fog computing, and edge computing in IoT systems

Features	Cloud computing	Fog computing	Edge computing
Network access	WLAN	Mostly LAN	LAN/WLAN
Mobility support	Low	Low	High
Scalability of servers	Very low	Distributed and low	Low or very low, depending on layers
Computing power location	Centralized	Distributed	Widely distributed
Reliability		High	Low
Deployment cost	Very high	Very low	Low
Maintenance	Cloud experts	Experts of enterprise	Expert, engineers, and employees of enterprise
Standardization	Yes	Yes	No
Operation mode	Standalone	Standalone/edge-cloud cooperation	Fog-cloud cooperation
Service coverage	Global	Local	Local
Service coverage	High	Low	Low
Virtualization	VM and containers	VM and containers	VM and containers
Constitutive elements	IoT devices, cloud servers	IoT devices, edge servers, cloud servers	IoT devices, fog nodes, cloud servers

applications (i.e., smart industry, smart transportation, smart healthcare) which increase the number of security vulnerabilities threatening the overall system security. The application-specific security techniques often demonstrated inadequate or/and impractical to satisfy the requirements of other applications.

- IoT systems are greatly heterogeneous in nature because of the diversity of involved devices, platforms, protocols, and they typically lack for standardization. These characteristics turn out to be obstacles that inhibit the development of efficient and comprehensive security solutions.

- IoT system captures a useful data that need to be analyzed to comprehend the behaviors of persons and their day-to-day events. Therefore, policymakers should exploit similar information to adapt their goods quickly and comply with personal inclinations and needs. Nevertheless, this result can turn IoT devices into spying devices to leak or seize personal information.

- Physical incidents in IoT systems are prone to increase due to fact that the majority of physical things can be ubiquitously and substantially accessible. Similarly, physical threats may be reasoned by unintentional destruction from natural disasters (i.e., floods or earthquakes) or catastrophes triggered by people (i.e., wars). Therefore, an active security solution should be environment-aware and take into account similar traits of IoT systems.

- The boundaries of IoT systems are hard if not impossible to define as they are continuously adapted according to the newly included devices owing to the

mobility of customers. This nature makes the IoT system grow constantly thereby enlarge potential attack surfaces and present more vulnerabilities.
- The progress of IoT systems from cloud computing, fog computing, to edge computing, along with main differences and similarities. Also, the chapter discusses the main advantage and disadvantages of each of them from a security perspective.

Consequently, intelligent learning algorithms are of great importance to fully comprehend and gain valuable knowledge about the normal and abnormal behaviors of things and other IoT components contained by this form of wide-scale systems. Nevertheless, deep learning approaches have the ability to foresee the anticipated system behaviors by learning inherent representations from past IoT traffic samples. Therefore, applying deep learning approaches can substantially improve the security mechanisms by renovating the security solutions of the IoT environments from merely enabling secure communications among IoT devices toward intelligent security systems.

References

1. B. Omoniwa, R. Hussain, M.A. Javed, S.H. Bouk, S.A. Malik, Fog/edge computing-based IoT (FECIoT): architecture, applications, and research issues. IEEE Internet Things J. (2019). https://doi.org/10.1109/JIOT.2018.2875544
2. M.H. Rehmani, A. Davy, B. Jennings, C. Assi, Software defined networks-based smart grid communication: a comprehensive survey. IEEE Commun. Surv. Tutorials (2019). https://doi.org/10.1109/COMST.2019.2908266
3. M.A. Al-Garadi, A. Mohamed, A.K. Al-Ali, X. Du, I. Ali, M. Guizani, A survey of machine and deep learning methods for internet of things (IoT) security. IEEE Commun. Surv. Tutorials (2020). https://doi.org/10.1109/COMST.2020.2988293
4. N. Chaabouni, M. Mosbah, A. Zemmari, C. Sauvignac, P. Faruki, Network intrusion detection for IoT security based on learning techniques. IEEE Commun. Surv. Tutorials (2019). https://doi.org/10.1109/COMST.2019.2896380
5. J. Lin, W. Yu, N. Zhang, X. Yang, H. Zhang, W. Zhao, A survey on internet of things: architecture, enabling technologies, security and privacy, and applications. IEEE Internet Things J. (2017). https://doi.org/10.1109/JIOT.2017.2683200
6. L. Atzori, A. Iera, G. Morabito, M. Nitti, The social internet of things (SIoT)—when social networks meet the internet of things: concept, architecture and network characterization. Comput. Netw. (2012). https://doi.org/10.1016/j.comnet.2012.07.010
7. I. Makhdoom, M. Abolhasan, J. Lipman, R.P. Liu, W. Ni, Anatomy of threats to the internet of things. IEEE Commun. Surv. Tutorials (2019). https://doi.org/10.1109/COMST.2018.2874978
8. H. Freeman, T. Zhang, The emerging era of fog computing and networking [The President's Page]. IEEE Commun. Mag. (2016). https://doi.org/10.1109/MCOM.2016.7497757
9. L. Hu et al., Cooperative jamming for physical layer security enhancement in internet of things. IEEE Internet Things J. (2018). https://doi.org/10.1109/JIOT.2017.2778185
10. Z. Wei, C. Masouros, F. Liu, S. Chatzinotas, B. Ottersten, Energy-and cost-efficient physical layer security in the era of IoT: the role of interference. IEEE Commun. Mag. (2020). https://doi.org/10.1109/MCOM.001.1900716
11. M. Hammoudeh, M. Arioua, Sensors and actuators in smart cities. J. Sens. Actuator Netw. (2018). https://doi.org/10.3390/jsan7010008

12. N. Wang, P. Wang, A. Alipour-Fanid, L. Jiao, K. Zeng, Physical-layer security of 5G wireless networks for IoT: challenges and opportunities. IEEE Internet Things J. (2019). https://doi.org/10.1109/JIOT.2019.2927379
13. N. Chen, M. Wang, N. Zhang, X. Shen, Energy and information management of electric vehicular network: a survey. IEEE Commun. Surv. Tutorials (2020). https://doi.org/10.1109/COMST.2020.2982118
14. G. Thompson, Ethernet: from office to data center to IoT. Comput. (Long. Beach. Calif.) (2019). https://doi.org/10.1109/mc.2019.2930099
15. X. Wu, M.D. Soltani, L. Zhou, M. Safari, H. Haas, Hybrid LiFi and WiFi networks: a survey. IEEE Commun. Surv. Tutorials (2021). https://doi.org/10.1109/COMST.2021.3058296
16. K.H. Chang, Bluetooth: a viable solution for IoT? [Industry perspectives]. IEEE Wirel. Commun. (2014). https://doi.org/10.1109/MWC.2014.7000963
17. M. Collotta, G. Pau, T. Talty, O.K. Tonguz, Bluetooth 5: a concrete step forward toward the IoT. IEEE Commun. Mag. (2018). https://doi.org/10.1109/MCOM.2018.1700053
18. G. Veerendra Nath, P. Kumar, K. Nageswara Rao, E. Nalin, G. Harshitha, K. Koteswara Rao, Ultra wide band multiple input multiple output antenna for internet of things applications. J. Comput. Theor. Nanosci. (2020). https://doi.org/10.1166/jctn.2020.8872
19. A. Attaran, R. Rashidzadeh, Chipless radio frequency identification tag for IoT applications. IEEE Internet Things J. (2016). https://doi.org/10.1109/JIOT.2016.2589928
20. L. Babun, H. Aksu, L. Ryan, K. Akkaya, E.S. Bentley, A.S. Uluagac, Z-IoT: passive device-class fingerprinting of ZigBee and Z-wave IoT devices (2020). https://doi.org/10.1109/ICC40277.2020.9149285
21. S.G. Varghese, C.P. Kurian, V.I. George, A. John, V. Nayak, A. Upadhyay, Comparative study of zigBee topologies for IoT-based lighting automation. IET Wirel. Sens. Syst. (2019). https://doi.org/10.1049/iet-wss.2018.5065
22. Z. Cao et al., Near-field communication sensors. Sensors (Switzerland) (2019). https://doi.org/10.3390/s19183947
23. V. Coskun, B. Ozdenizci, K. Ok, The survey on near field communication. Sensors (Switzerland) (2015). https://doi.org/10.3390/s150613348
24. R. Want, Near field communication. IEEE Pervasive Comput. (2011). https://doi.org/10.1109/MPRV.2011.55
25. O. Georgiou, U. Raza, Low power wide area network analysis: Can LoRa scale? IEEE Wirel. Commun. Lett. (2017). https://doi.org/10.1109/LWC.2016.2647247
26. A.J. Onumanyi, A.M. Abu-Mahfouz, G.P. Hancke, Cognitive radio in low power wide area network for IoT applications: recent approaches, benefits and challenges. IEEE Trans. Ind. Inform. (2020). https://doi.org/10.1109/TII.2019.2956507
27. H. Ruotsalainen, J. Zhang, S. Grebeniuk, Experimental investigation on wireless key generation for low-power wide-area networks. IEEE Internet Things J. (2020). https://doi.org/10.1109/JIOT.2019.2946919
28. S. Jaffry, R. Hussain, X. Gui, S.F. Hasan, A comprehensive survey on moving networks. IEEE Commun. Surv. Tutorials (2021). https://doi.org/10.1109/COMST.2020.3029005
29. R. Borralho, A. Mohamed, A. Quddus, P. Vieira, R. Tafazolli, A survey on coverage enhancement in cellular networks: challenges and solutions for future deployments. IEEE Commun. Surv. Tutorials (2021). https://doi.org/10.1109/COMST.2021.3053464
30. G. Pardo-Castellote, OMG data-distribution service: architectural overview, (2003). https://doi.org/10.1109/ICDCSW.2003.1203555
31. F. De Rango, G. Potrino, M. Tropea, P. Fazio, Energy-aware dynamic internet of things security system based on elliptic curve cryptography and message queue telemetry transport protocol for mitigating replay attacks. Pervasive Mob. Comput. (2020). https://doi.org/10.1016/j.pmcj.2019.101105
32. S.R. Akbar, K. Amron, H. Mulya, S. Hanifah, Message queue telemetry transport protocols implementation for wireless sensor networks communication—a performance review (2018). https://doi.org/10.1109/SIET.2017.8304118

33. S. Vinoski, Advanced message queuing protocol. IEEE Internet Comput. (2006). https://doi.org/10.1109/MIC.2006.116
34. R. Godfrey, D. Ingham, R. Schloming, OASIS advanced message queuing protocol (AMQP) version 1.0. OASIS Stand. (2012)
35. Z. Shelby, K. Hartke, C. Bormann, The constrained application protocol (CoAP), Rfc 7252 (2014)
36. C. Bormann, S. Lemay, H. Tschofenig, K. Hartke, B. Silverajan, E.B. Raymor, CoAP (Constrained Application Protocol) over TCP, TLS, and WebSockets. RFC 8323 (2018)
37. C. Xie, B. Yu, Z. Zeng, Y. Yang, Q. Liu, Multilayer internet-of-things middleware based on knowledge graph. IEEE Internet Things J. (2021). https://doi.org/10.1109/JIOT.2020.3019707
38. A.J. Brush, J. Albrecht, R. Miller, J. Albrecht, A.J. Brush, M. Hazas, Smart Homes. IEEE Pervasive Comput. (2020). https://doi.org/10.1109/MPRV.2020.2977739
39. A. Ghubaish, T. Salman, M. Zolanvari, D. Unal, A.K. Al-Ali, R. Jain, Recent advances in the internet of medical things (IoMT) systems security. IEEE Internet Things J. (2020). https://doi.org/10.1109/JIOT.2020.3045653
40. Y.A. Qadri, A. Nauman, Y. Bin Zikria, A.V. Vasilakos, S.W. Kim, The future of healthcare internet of things: a survey of emerging technologies. IEEE Commun. Surv. Tutorials (2020). https://doi.org/10.1109/COMST.2020.2973314
41. H. Habibzadeh, K. Dinesh, O. Rajabi Shishvan, A. Boggio-Dandry, G. Sharma, T. Soyata, A survey of healthcare internet of things (HIoT): a clinical perspective. IEEE Internet Things J. (2020). https://doi.org/10.1109/JIOT.2019.2946359
42. M.N. Bhuiyan, D.M.M. Rahman, M.M. Billah, D. Saha, Internet of Things (IoT): a review of its enabling technologies in healthcare applications, standards protocols, security and market opportunities. IEEE Internet Things J. (2021). https://doi.org/10.1109/JIOT.2021.3062630
43. H. Wang et al., Architectural design alternatives based on cloud/edge/fog computing for connected vehicles. IEEE Commun. Surv. Tutorials (2020). https://doi.org/10.1109/COMST.2020.3020854
44. S. Gyawali, S. Xu, Y. Qian, R.Q. Hu, Challenges and solutions for cellular based V2X communications. IEEE Commun. Surv. Tutorials (2021). https://doi.org/10.1109/COMST.2020.3029723
45. P. Kumar, Y. Lin, G. Bai, A. Paverd, J.S. Dong, A. Martin, Smart grid metering networks: a survey on security, privacy and open research issues. IEEE Commun. Surv. Tutorials (2019). https://doi.org/10.1109/COMST.2019.2899354
46. A. Ghosal, M. Conti, Key management systems for smart grid advanced metering infrastructure: a survey. IEEE Commun. Surv. Tutorials (2019). https://doi.org/10.1109/COMST.2019.2907650
47. K. Tange, M. De Donno, X. Fafoutis, N. Dragoni, A systematic survey of industrial internet of things security: requirements and fog computing opportunities. IEEE Commun. Surv. Tutorials (2020). https://doi.org/10.1109/COMST.2020.3011208
48. T. Qiu, J. Chi, X. Zhou, Z. Ning, M. Atiquzzaman, D.O. Wu, Edge computing in industrial internet of things: architecture, advances and challenges. IEEE Commun. Surv. Tutorials (2020). https://doi.org/10.1109/COMST.2020.3009103
49. N. Magaia, R. Fonseca, K. Muhammad, A.H.F.N. Segundo, A.V. Lira Neto, V.H.C. De Albuquerque, Industrial internet-of-things security enhanced with deep learning approaches for smart cities. IEEE Internet Things J. (2021). https://doi.org/10.1109/JIOT.2020.3042174
50. G. Manogaran, C.H. Hsu, B.S. Rawal, B. Muthu, C.X. Mavromoustakis, G. Mastorakis, ISOF: information scheduling and optimization framework for improving the performance of agriculture systems aided by industry 4.0. IEEE Internet Things J. (2021). https://doi.org/10.1109/JIOT.2020.3045479
51. N.N. Misra, Y. Dixit, A. Al-Mallahi, M.S. Bhullar, R. Upadhyay, A. Martynenko, IoT, big data and artificial intelligence in agriculture and food industry. IEEE Internet Things J. (2020). https://doi.org/10.1109/jiot.2020.2998584
52. V. Hassija, V. Chamola, V. Gupta, S. Jain, N. Guizani, A survey on supply chain security: application areas, security threats, and solution architectures. IEEE Internet Things J. (2021). https://doi.org/10.1109/JIOT.2020.3025775

53. M. Asante, G. Epiphaniou, C. Maple, H. Al-Khateeb, M. Bottarelli, K.Z. Ghafoor, Distributed ledger technologies in supply chain security management: a comprehensive survey. IEEE Trans. Eng. Manag. (2021). https://doi.org/10.1109/TEM.2021.3053655
54. M. Razaghi, M. Finger, Smart governance for smart cities. Proc. IEEE (2018). https://doi.org/10.1109/JPROC.2018.2807784
55. M. Sookhak, H. Tang, Y. He, F.R. Yu, Security and privacy of smart cities: a survey, research issues and challenges. IEEE Commun. Surv. Tutorials (2019). https://doi.org/10.1109/COMST.2018.2867288
56. R. A. MARKET, Global Cloud Computing Market 2021–2025 (2021). (Online). Available: https://www.researchandmarkets.com/reports/5316719/global-cloud-computing-market-2021-2025?utm_source=BW&utm_medium=PressRelease&utm_code=lshczn&utm_campaign=1531847+-+Global+Cloud+Computing+Market+(2021+to+2025)+-+Featuring+Adobe%2C+Alphabet+and+Amazon+A
57. A. Masood, D.S. Lakew, S. Cho, Security and privacy challenges in connected vehicular cloud computing. IEEE Commun. Surv. Tutorials (2020). https://doi.org/10.1109/COMST.2020.3012961
58. K. Gai, J. Guo, L. Zhu, S. Yu, Blockchain meets cloud computing: a survey. IEEE Commun. Surv. Tutorials (2020). https://doi.org/10.1109/COMST.2020.2989392
59. F. Bonomi, R. Milito, J. Zhu, S. Addepalli, Fog computing and its role in the internet of things (2012). https://doi.org/10.1145/2342509.2342513
60. IEEE Communications Society, IEEE Std 1934–2018: IEEE Standard for Adoption of OpenFog Reference Architecture for Fog Computing (2018)
61. P. Singh, A. Kaur, G.S. Aujla, R.S. Batth, S. Kanhere, DAAS: dew computing as a service for intelligent intrusion detection in edge-of-things ecosystem. IEEE Internet Things J. (2020). https://doi.org/10.1109/JIOT.2020.3029248

Chapter 3
Internet of Things Security Requirements, Threats, Attacks, and Countermeasures

This chapter elaborates on different security aspects to be taken into accounts during the development and the deployments of IoT architecture. To make the reader about the security of the IoT based system, this chapter begins by defining the contemporary security requirements that should be considered to realize a reliable and trustworthy IoT environment. Then, the discussion extends to differentiate different concepts of IoT security i.e., threat, vulnerability, countermeasure, attacks, risks; and also explain how the concepts relate to each other. Later, a systematic taxonomy is presented for classifying IoT attacks according to IoT assets, where each class of IoT is further classified into more subcategories. Finally, the discussion of each elaborate different categories of IoT attack indicating their main security targets and possible IoT countermeasures.

To sum up, this chapter intends to provide a comprehensive overview regarding IoT security vulnerabilities, threats, countermeasures, risks along with practices of handling them all through the following sections:

- Security Requirements in Internet of Things
- IoT threats, Attacks, vulnerabilities, and risks
- Today's IoT attacks and Countermeasures
- IoT attack surfaces
- Summary and Learnt Lessons.

3.1 Security Requirements in Internet of Things

The concept of information assurance can be identified as the practice of guaranteeing that information systems will function as required when required, while

Electronic Supplementary Material The online version of this chapter (https://doi.org/10.1007/978-3-030-89025-4_3) contains supplementary material, which is available to authorized users.

M. Abdel-Basset et al., *Deep Learning Techniques for IoT Security and Privacy*, Studies in Computational Intelligence 997, https://doi.org/10.1007/978-3-030-89025-4_3

keeping them secure and protected. As reported by the National Institute of Standards and Technology [1, 2], information assurance is described as *"Measures that protect and defend information and information systems by ensuring their availability, integrity, authentication, confidentiality, and non-repudiation. These measures include providing for restoration of information systems by incorporating protection, detection, and reaction capabilities"*. These five mainstays of information assurance are applicable as security requirements of IoT-based systems for the reason that it combines digital information world with a physical counterpart, data resources, and communication networks [3]. However, with the continuous advancements in cyber and physical aspects of IoT paradigms (i.e., fog/edge computing, fog-cloud interaction, edge-cloud interactions, etc.), these five mainstays of information assurance became insufficient to thoroughly reflect the overall security requirements or goals of IoT environments (see Fig. 3.1).

IoT combines the physical and Internet-connected world to deliver a smart collaboration between the physical entities and the enclosing environments (i.e., smart buildings, companies, etc.). Usually, IoT devices operate in different environments to achieve various objectives. Nevertheless, their operation ought to encounter a thorough cyber security requirements and the physical security requirements [4]. The composite nature of IoT environments stem from the involvement of multidisciplinary elements, networks, computations, etc. This further broaden the attack surfaces of IoT-based system and make the task of satisfying the security constraints more challenging. To fulfill the anticipated IoT security requirement mostly necessitate a solution with all-inclusive considerations. Nevertheless, IoT devices generally operate in a crowded and open environment. Therefore, IoT devices are prone to be physically accessed by attackers/intrudes. IoT devices are typically interconnected across wireless communication networks in which attackers/intruders could imitate eavesdropping to uncover secret information from the communication. The resource-constrained nature of IoT devices makes them unable to support complicated security

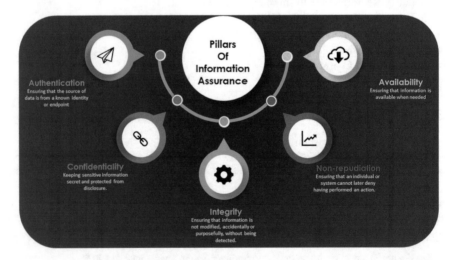

Fig. 3.1 The classical pillars of information assurance

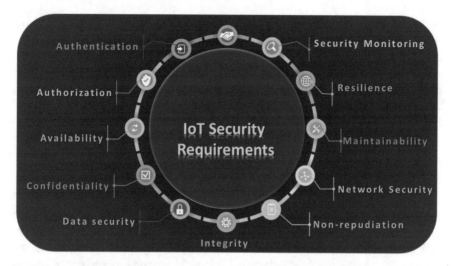

Fig. 3.2 The primary security requirements for cyber-physical IoT environments

solutions [5]. Therefore, maintaining the security or privacy of IoT-based systems is a multifaceted and challenging task that attract much interest in both academic and industrial domain. Provided that the major purpose of the IoT-based system is to permit easy access to anyone, anywhere and anytime, attack surfaces turn out to be more prone to various attacks [6, 7].

The information assurance can be deemed as high-level and abstract classes of security goals or requirements just for different digital information systems. Therefore, this section presents detailed discussions regarding the IoT security requirements and/or goals. In addition, the reasons that make these requirements are challenging to fulfill in the context of Industry 4.0 applications, thereby delivering the readers with valuable insights necessary to realize the reasons for which the debated security conditions are challenging to satisfy using the standard approaches.

Concisely, the security requirements (see Fig. 3.2) of an IoT-based system are as follows.

3.1.1 Authentication (S1)

The identity of entities should be absolutely determined before gaining access to data or executing any tasks. Nevertheless, owing to the evolving characteristics of IoT-based systems, the requirement for authenticating different entities vary between systems, applications, and layers. For instance, IoT-based systems necessitate robust authentication schemes, in which services and applications should afford great security even at the expense of other factors i.e., elasticity. Trade-offs are a key challenge facing the design of an efficient authentication system. A typical example of that is the trade-off between the safety and security of IoT healthcare devices, where the design of an authentication scheme necessitates a better balance between safety and security. Likewise, the trade-offs in IoT-based may include security-privacy trade-off, security-computation trade-off, security-battery trade-off, and so many others depending on the underlying system conditions [8]. Thus, the design of efficient authentication scheme for IoT-based system often necessitate large consideration for harmonization of system restraints and deliver strong security systems [9].

3.1.2 Authorization (S2)

Authorization involves granting users access privileges to different IoT elements i.e., IoT devices, software, applications, etc. The term "users" here refers to humans, services, machines, and other forms of cyber-physical operators. For instance, the sensory aggregated data have to be provided to and accessed by authorized entities and clients [10]. Which means, a task ought to be achieved only when the requesting users have acceptable authorization to have control over it. The central challenge of authorization in IoT-based systems is a way to effectively grant access in an environment not consisting of human individuals only, but also include physical things, that need to be authorized to work together with other IoT entities. Besides, to cope with massive volumes of data in the heterogeneous ecosystem, the data be required to be safeguarded all through the sensing and communication procedure and have to be rendered available merely to approved entities [11].

3.1.3 Availability (S3)

The availability requirement implies that the authorized entities always have access to their data and/or services in IoT-based system. Availability is an essential property of the effective deployment of various IoT solutions. However, IoT devices, services, and systems are prone to be portrayed unreachable by a variety of threats, i.e., active jamming or Daniel of service attacks. Consequently, guaranteeing the constant accessibility of IoT applications and/or services for the clients can be considered as a crucial characteristic of any reliable IoT-based system [12].

3.1.4 Confidentiality (S4)

Confidentiality is known as a critical security trait of IoT-based systems, wherein a variety of information with different degrees of sensitivity can be stored on or transmitted between local devices, edge, fog, or cloud servers, which must not be exposed or accessed by unauthorized entities. For example, personal information, patients' medical records, secret economical information, military data are extremely confidential and ought to be protected against intrusions and unapproved access. Nevertheless, under a circumstance like with healthcare IoT devices, although data are confidentially put in storage and transmitted as well as encrypted communications, the attackers could stay detecting the presence of the physical things and could yet trace the owner. Hence, the confidentiality of the location of owners or communication channel is revealed and becomes in danger [13].

3.1.5 Data Security (S5)

Henceforward, the digital world record maintaining the security of data as an important security aspect, and the introduction of IoT make the data security indispensable aspect of the design of secure IoT systems. Several studies considered the data confidentiality as security necessity for IoT data [14–18]. Conventionally nevertheless, data integrity and data availability are deemed more advantageous than confidentiality especially in industrial ecosystems [19, 20], thanks to the fact that they have a quantifiable commercial effect. This is an inappropriate point of view in the context of an environment of networked devices, and the companies rapidly trending to shift their offline systems to be internet-connected systems.

According to companies-based survey studies [16], it was discovered and demonstrated that the security of data is an essential factor motivating the companies to migrate to Industry 4.0 [18]. In addition, it was stated that companies were cautious to take on data-sharing-based techniques (i.e., prevention, cloud sharing, fault detection, etc.) because of the non-existence of evidence about the safety or security of these techniques during intellectual property protection. Consequently, this indicated the requirement for a consistent approach to protect rational estate of the existence of data sharing procedures. In the early stages, a pervasive opinion shares those enterprises are having hesitation toward the dependence on cloud servers for storing and sharing IoT data [21]. Nevertheless, it is also observed that the majority of IoT data violations occur within companies, rather than at cloud providers. Then, a cloud-based storage was introduced for narrowing the surface of attacks at the company side as well as the cloud side. However, data loss alleviation is emerged as an additional necessity, pinpointing four crucial procedures necessary for developing an effective solution. These factors include identification, prevention, documentation, and notification.

The challenges in this field are associated with three interfering aspects: Firstly, the resource-constrained nature and mobility of IoT devices, data security techniques have to function in such a way the enable very limited resource consumption. Secondly, the need for data security becomes extremely high especially in the case of sensitive IoT applications or services. Thirdly, numerous IoT facilities are supported through data sharing, but in data-sensitive environments, confidentiality is of paramount interest, which often imposes many problems.

The field of data security span multiple subfields rather than just encryption methods. The convergence of IoT and industry 4.0 paradigm bring us many benefits, which enable better exploitation of existing IoT data. This certainly includes data sharing and other data-dependent operations, that might be found at any piece of the system and even beyond the organization's limits. While encryption algorithms provide a mean to support preferential data sharing, this subsection elaborates the other strategies for keeping the confidentiality of IoT data.

Data transmission: Generally, the MQTT is a commonly employed data sharing protocol between different entities in IoT-based systems. However, it just endorses user/password authentication and does not afford any security procedures through the network or application layer. This turns out to be challenging particularly in critical IoT applications. As a remedy, Transport Layer Security (TLS) was introduced to deliver a protected layer that MQTT could securely operate upon [22]. However, this solution presents a significant computational overhead for the IoT edge devices (either stationary or mobile) that are currently should supervise TLS contexts. Thus, make TLS an impractical choice for the environment having very limited resources such as wireless sensor networks. A trustworthy hardware expansion can be employed at the edge devices to enable storing encryption keys and assisting the TLS context management. While contemporary devices could have low-cost dependable hardware, this is not always conceivable for all IoT devices.

Instead, the before mentioned issues can be tackled with innovative protocols like Open Platform Communications Unified Architecture (OPC UA) [23] which support authentication, encryption, and hardware acceleration for the cryptographic elementary, making it a suitable choice for real-time and lightweight IoT solutions. Modern empirical deployment integrates this with trustworthy hardware to enable reliable connections but admits that extra investigation is essential in either academia or industry. After All, it is notable that irrespective of the applied security scheme, from a power and efficacy perspective, it is recommended to apply selective encryption for the data packets that could impair the system when tinkered with [24].

Outdoor parties: the confidentiality of IoT data in either storage or transmission state, is habitually achieved via cryptographic techniques. The challenges in determining appropriate ciphers for the extremely different IoT ecosystem are revolving around energy concerns and other resource limitations. Regardless of the employed cipher, the process of key distribution and management remains a challenging task. Other common challenges are presented in the context of Cloud-assisted and Fog-assisted IoT environments including trustworthy storage, effective search, appropriate usage, and dependable data removal. Since the Cloud plentiful computing and storage resources, connectivity to the cloud is usually essential for large-scale IoT applications. Having an appropriate encryption technique, the IoT data can be easily stored and kept secure at the cloud datacenters [25], however, once stored, it is impractical to use it for any purpose except for restoring it to be decrypted. To resolve this, numerous academic and industrial efforts are being devoted to developing new cryptographic methods such as homomorphic encryption, which enables the use of encrypted data in computational tasks, and searchable encryption, which promotes searching within encrypted data. In another way, an anonymous aggregate encryption scheme was introduced for IoT data sharing, in which IoT devices are authorized to encrypt data into a single ciphertext, which is later decrypted with many receivers using their personal keys while maintaining their respective anonymity.

Distributed IoT devices communicate and exchange the locally stored data with different servers responsible for data aggregation, elimination of redundant records, and data indexing and encryption to be later stored at remote cloud data centers. Follow, Customers could search the stored data via trapdoor requests, indicating that the search activity is applicable to the encrypted data. To obtain the queried data, the customers download the encrypted findings, which are later decrypted using privileged keys of customers. Thus, the cloud backend unable to access any decrypted data. however, the introduction of big data analytics on the cloud makes this solution impractical.

Data privacy: privacy and proprietorship of data are becoming a crucial matter for almost all enterprises and even governments. As a result of the pervasive acceptance of cloud storage and cloud services, privacy issues are greatly increased and gained increased attention in both academia and industry [26]. Thanks to a large volume of data generated and acquired via the physical entities, it is easy to build thorough profiles for each user, leaving the privacy of user's information at great threat [27]. In the stream of solutions to address this, an anonymous data aggregation scheme can be employed. In addition, the latest statute of the General Data Protection Regulation (GDPR) [28] necessitates that data privacy considerations must be taken strictly by manufacturers of different IoT devices, which is known as privacy-by-design. This consideration ought to be contemplated during the design of any IoT application that intends to interact with any kind of private or sensitive information i.e., military information, governmental data, human's personal information, etc. Privacy-preservation did not just relate to the aggregation of data at cloud; however, it also involves the mystification or exclusion of metadata and further characteristics that could be leveraged by attackers/intruders. For instance, in wireless sensor networks, physical devices are geographically distributed across broad range of locations, where an adversary may possibly try to find the source device of particular IoT traffic flow according to the information flow. As a solution to this, a method for preserving the privacy of the source location have to be employed [29].

3.1.6 Integrity (S6)

The sensed data in IoT environments are normally exchanged between different parties over wireless networks and only the authorized entities are permitted to adjust. Integrity aspects are therefore essential for guaranteeing an efficient verification method to discover any data adjustment through the transmissions across vulnerable wireless communications. Integrity properties can enable securing the IoT-based system from malevolent entries that could initiate structured query language (SQL) injection attacks. A defect in integrity scrutiny might permit for illegal adjustment of the memory-saved information in physical IoT devices, thereby influencing the majority of the operations of the IoT entities for a prolonged period while being identified. IoT-based applications often exhibit different set integrity conditions. For instance, the implantable healthcare devices necessitate active integrity inspection counter to arbitrary errors for the reason that they explicitly influence patients' lives. Failure, mistakes, or changes of information in some conditions can result in sacrificing human lives [30].

3.1.7 Non-repudiation (S7)

Non-repudiation is a security mainstay designed to deliver access logs that can be used as proof under circumstances where individuals and things cannot deny a particular action. Mostly, non-repudiation still not contemplated a major security requirement in different IoT ecosystems [5]. Nevertheless, non-repudiation could be viewed as a crucial security property in certain situations in which two involved entities unable to repudiate the undergoing operation i.e., banking systems, payment system, marketing system [31, 32].

3.1.8 Network Security (S8)

Realizing acceptable network security involves numerous issues i.e., dependable routing, authentication, secure transport, etc. some of them already addressed by the aforementioned security requirements, hence this therefore emphasis the security of network infrastructure. Thanks to the increased complexity of IoT networks as the result of enlargement in the number and scale of connected devices, it becomes problematic to offer a security solution or protocol that can cope with this fast expansion. Therefore, numerous scalability and performance matters necessary to be tackled, including throughput, bandwidth, delay, etc. To this end, traffic control, network configurations, and security systems depend on commercial software making incorporation in standard management schemes impractical. Simultaneously, the network infrastructure should be accommodating to cope with vibrant ecosystems. In an attempt to address these issues, two networking models emerged to enable isolating the control and configuration tasks from data transmission making networks agile and flexible, namely, Software Defined Networking (SDN) and Network Function Virtualization (NFV). SDN emphasis management and configuration of network service, whilst NFV concentrates on virtualized settings to implement policies of the IOT network and run detection or defense operations on an abstracted layer without considering the location of these functions and the way they are used to route the traffic [33, 34].

3.1.9 Maintainability (S9)

Maintainability is one of the IoT security goals that concern the process of configuring, reconfiguring, and upgrading different components of IoT-based systems. With the emergence of Industry 4.0, the concept of maintainability turns out to be critical as the system configuration and software in IoT systems ought to have the capacity to be upgraded to offer protection against formerly unidentified security threats. Upgradability could be deemed as a countermeasure counter to a different form of IoT threats, since it enables constant modifications to firewall configurations according to the emergent threats, in addition to software pieces for recently encountered software vulnerabilities. As later discussed, the primary challenges facing maintenance are associated with resource limitations and the dynamic nature of IoT ecosystems, rendering the conventional preservation approaches to be impractical to effectively satisfy the requirements in this realm [5, 6].

3.1.10 Resilience (S10)

In 2016, an IoT security framework [35] introduced by the Industrial Internet Consortium (IIC) describes resilience as *"the emergent property of a system that behaves in a manner to avoid, absorb and manage dynamic adversarial conditions while completing the assigned missions, and reconstitute the operational capabilities after causalities"*. This definition combines multiple facets of system security like safety and dependability. In another way, the authors of [3] considered resilience and safety as a tightly related concept in the cyber-physical IoT environment, where resilience was identified as *"A resilient control system is one that maintains state awareness and an accepted level of operational normalcy in response to disturbances, including threats of an unexpected and malicious nature."*; while safety was defined *"The condition of being safe from undergoing or causing hurt, injury, or loss."* In Fact, [36] and [24] characterize resilience as a critical security challenge for IoT-based systems. The impacts of the resilience obligations on the security realm necessitate mechanisms that ought to deliver the ability to keep on regular system functions when portions of the system are deemed to malfunction. This might be accomplished by deflecting tasks to alternative efficient IoT elements, or via other methods, habitually belonging to one of three established approaches for improving the security and resilience of IoT-based systems, namely diversity, redundancy, and hardening [37].

- Diversity: security approach for designing IoT elements employing a variety of element categories (i.e., different software and hardware implementations) so that vulnerabilities belonging to a particular category of IoT components have a partial effect on the IoT-based system.
- Redundancy: security approach for installing extra redundant IoT elements in a system in such a way that enables the system to maintain standard or almost satisfactory operations even when some IoT elements are compromised.

- Hardening: security approach for boosting specific IoT elements or categories of IoT elements, such as firewalls and tamper-resistant hardware, in such a way that makes them more difficult to be impaired or compromised.

The way this requirement must be fulfilled greatly relies on the underlying scenario. On one hand, in Wireless Sensor Networks, the redundancy of the sensors is likely to present, implying that a few numbers of a compromised set of sensors might be remained enclosed and the corresponding output abandoned while waiting for the issue to be handled. On the other hand, in a power station, it could be disastrous to completely deactivate one generator when some of its modules were in a compromised state. As An Alternative, the functionality of compromised modules should be provided with some other means, or momentarily use another generator to produce the required energy and ensure accomplishing the same level of processes [38, 39].

3.1.11   Security Monitoring (S11)

Dynamic monitoring of the conduct of an IoT system is regarded as an efficient strategy for early identification and reaction to malicious actions, and the systems affording these facilities are generally recognized as Intrusion Detection Systems (IDSs). Thus, the security of IoT-based system could be further improved using two vital factors: first, the capability of monitoring underlying IoT infrastructure; second, the ability to effectively act in response to well-known and anonymous threats [40]. The motivation these factors are regarded chiefly vital for secure IoT stems from the truth that legacy, partially protected devices, and protocols (i.e., Modbus, MQTT) are expected to be also involved in IoT-based systems, which are unable to continuously be patched to protect against established security susceptibilities, and hence necessitate constant monitoring. Another motivation could be presented in enhancing congestion management generally while delivering defense against Denial of Service (DoS) attacks [41].

A variety of IDSs has been developed to operate in general information technology, which can be categorized into two main kinds of namely signature-centered IDSs and anomaly-centered IDSs. The signature-centered IDSs seeks to distinguish malevolent intrusions by exploring the relationship between the new pattern and the early learnt signatures of well-known outbreaks. Nevertheless, this methodology makes the signature-based IDS fails to distinguish newly discovered categories of attacks. Also, the complexity and the burdens of these approaches grows up with the expansion of the number of recently identified attack patterns, which mean larger amount of signatures, which in turn weaken the efficiency of these approaches [42]. Moreover, signature-based IDS regularly requires involvement of human professionals for scrutinizing and analyzing the signatures of emerging intrusions. In contrast, anomaly-based IDS enable differentiating unknown intrusions broadly occur in nearly all IoT ecosystems. It is worth noting that Latency is the main challenge for monitoring the security of IoT-based systems, because of the remotely distributed devices over IoT networks. Hence, high network latency can be presented as a result of time taken by the satisfactory reaction to intrusions, mainly when cloud security solutions are employed. Another emergent challenge can be the large imbalance of threat data. in other words, scarce and rarely occurring intrusions can be difficult to be detected as the IDSs usually have little information about such threats. With the proliferation of IoT technologies, it turns out to be difficult to monitor threats within large-scale and intense IoT ecosystems containing a very big number of heterogeneous entities, sensors, smart appliances, each with a different set of threats. Besides, since the IoT operations are mostly the consequence of automatic procedures, the traffic flows incline to be equitably stationary and periodic, challenging the design of robust IDS for efficient security monitoring [6].

Fulfilling an information security goal does not certainly necessitate that an organization must retain all the previously mentioned mainstays in position. Categorizing the data and Information is a complicated subject thanks to the fact that not all information is confidential or vital. Appropriate threat modeling of devices, the applications they host, and data obligate IoT-based systems to recognize the sensitivities of different data elements either separately or in combined form. Combination dangers of huge, apparently non-threatening IoT datasets impose many security challenges. Perfectly delineated data types and combinatory restraints allow certain security goals such as confidentiality or integrity to be identified for each class of data or multifaceted information.

3.2   IoT Threats, Attacks, Vulnerabilities, and Risks

A considerable amount of academic efforts have been dedicated to clearly define and discriminate for the concepts of threats, vulnerability, and risks [3]. In view of this, this section explores and provides an in-depth discussion of these concepts on context of the cyber-physical IoT systems.

3.2.1   IoT Threats

It is imperative to differentiate a threat from a threat actor. Typically, each threat is known to have a corresponding threat actor. For instance, when a thief is invading a certain house, it is enticing to believe the thief as the real threat, however, it is more precise and valuable to regard him as the threat actor seeking to attack the house for different malevolent targets. most remarkably his self-serving need to take the residents' appraised holdings. In view of this, the threat is essentially the potential for the theft to be done, or more normally signifies the exploit potential.

Threats can thus occur in two main forms, namely man-made threats, and natural threats. The man-made IoT threats involve overall information assurance threats to data, communication, control, and applications. Besides, the integration of physical devices makes the IoT-based system prone to physical threats, environmental threats, supply chain threats, software quality threats, and a lot of other threats. Cyber-physical IoT devices are subject to physical reliability and resilience threats beyond just the compromise and degradation of the computing platform. Other engineering specialties come into force in the cyber-physical systems, such as actuation devices, filters, controllers, sensor feedback, state evaluation and management, classical control theory, and others that use physical things to control and regulate physical system states. Moreover, the threats can also affect the state assessment filters, transfer affairs of the control system, and other internal control loop artifacts that likely have explicit reactions and effects on the physical infrastructure. On the other hand, hurricanes, floods, and Tornados is an example of natural threats, where threat actor is the planet's weather [4].

The security threats within the IoT environment could be divided into two broad categories namely cyber-threats and physical threats. Besides, the cyber-threats could be categorized into passive threats or active threats. The subsequent clauses offer an in-depth discussion of these different categories of IoT threats.

3.2.1.1   Cyber-Threats

Passive threats: Passive threats are defined as a kind of IoT threats that mostly target confidentiality of IoT system the in which, the intruder/attacker monitor the content of traffic flows or make a copy of the packets' content. Thanks to the passive threats there is no damage to the system. The highly essential thing to know about the passive threats is that the victims are not get notified about the attack/intrusion. In passive threats, an attacker/intruder watches a IoT system and network communications and searches for gaping points and other vulnerabilities. For instance, they may possibly make use of an unpatched system or exploit a terminated license at a security device. As Soon as the attackers/intruders have penetrated the IoT network, they became capable of collecting information within a pair of ways. In a foot-printing passive threat, the attackers/intruders seek to aggregate as far information as they can on the way to be utilized to attack the IoT infrastructure or network in a subsequent phase. A typical example of this, attackers that record IoT traffic employing a packet analyzer tool, to be used later for malicious purposes. Other examples can be the keylogger attacks, where attackers ambush the users to submit their credentials and logs them to be used for unethical purposes in later times.

There popular cases of passive threats can be defined as follow:

- **Traffic analysis**: an attacker/intruder supervises communication channels for aggregating a collection of information, involving human identities, machine specifications, location information, and kinds of applied encryptions used.
- **Release of message contents**: this refers to situations, where an attacker/intruder employs virus or malware to take on monitoring vulnerable messaging tools such

as unencrypted voice calls, email, etc.;—and seize to access and release their sensitive content.

- **Passive reconnaissance**: it represents the situation, where an attacker/intruder attempt to obtain valuable information regarding the target internet-connected cyber-physical system with no transmission of IoT traffic to the aimed node, network, or server. A typical example for this, an attacker/intruder who embraces browsing a website substance for pertinent information (e.g., contact information). Which could be exploited in active attacks or retrieving vulnerable files on the target side, such as financial reports, scholarly property, and industrial details.

Identifying passive threats is considered extremely challenging from a deep learning perspective for the reason that they do not include data modification in any manner.

Active threats: Active threats refer to the kind of IoT threats that mostly target the availability and the integrity of IoT systems, in which, the attacker/threats struggle to modify or change the content of IoT traffic. In other words, the threats involve a series of interferences, interruptions, and adjustments. Thanks to active threats, the IoT system, and the corresponding resources are under direct damage. The highly essential thing to know about the passive threats is that the victims are not get notified about the attack/intrusion.

In active threats, the attackers/intruders are restricted to eavesdrop on communication channels, however, they also apply modifications or changes to the settings of IoT-based systems, communication management, deny of services, and others. Active threats usually entail exploiting information acquired by passive threats to compromise different system elements or networks.

There popular categories of active threats can be defined as follow:

- **Interruption threats**: this is also recognized as a Masquerade attack, where the intruders/attackers might pretend to be a certain system entity to get into or to gain superior privileges than they are permitted. These kinds of attacks could be committed by means of leaked passwords and logons, finding software gaps, or discovering a route around the authentication procedure. They might be initiated with somebody from the organization itself or from outside when there is a connection to a public network.
- **Modification threats**: These kinds of attacks could be accomplished in two directions namely Modification of messages or Replay. Modification of messages refers to a situation in which some fraction of a message is changed, or the message's content undergoes some delay or rearrangement to generate an unauthorized impact. On the other hand, Replay refers to the situations where a chain of messages or events might be captured and resent by the attackers/intruders to create an authorized consequence.
- **Fabrication threats**: The threats that might lead to denial of service, where the attackers/intruders seek to thwart legal entities from gaining access to some system facilities or services. Which means that the attackers can access the actions while the authorized clients are banned from utilizing the services. One More form of

Table 3.1 Comparison between passive and active attacks

Features	Active threats	Passive threats
Modification	Modification in information takes place	While in passive attack, Modification in the information does not take place
Security target	Endanger the availability and Integrity	Endanger for Confidentiality
Community focus	attention is on detection	Attention is on prevention
The impact	System is always damaged	There is not any harm to the system
Victim awareness	A victim gets notified about the threats	A victim does not get notified about the threats
Resources	IoT resources can be changed	IoT resources are not changing
Services	They interrupt the services of the system	Just acquire the information and messages in the system
Categories	• Interruption threats • Modification threats • Fabrication threats	• Traffic analysis • Release of message contents • Passive reconnaissance

service denial is the disturbance of a complete network by either deactivating the network or by overburdening it with messages.

Different from passive threats, active threats are more likely to be detected promptly by the target once they are triggered. To sum up, a tabular comparison between active and passive threats is shown in Table 3.1.

3.2.1.2   Physical Threats

Physical threats represent the actions of causing physical damage to physical things of the IoT system either intentionally or unintentionally. In intentional physical threats, the attacker/intruder mostly does not have technological facilities to perform initiate any kind of cyber-threats. Hence, the attacker/intruder could merely influence the available physical things and related other elements of IoT that cause the services and/or applications to be terminated. With the rapid proliferation of IoT systems across different industries, the scale of physical attacks is continuously expanded as the result of increasing the number of reachable physical IoT objects everywhere. On the other hand, the physical threats might be initiated by unintentional destruction from disasters initiated by humans i.e., wars; or environmental catastrophes, including floods, earthquakes, or volcanoes [43–45].

In the literature, some people use the word threat and or attack interchangeably as they considered having the same meaning. On other hand, some others prefer to be specified and consider the two terms different. This book follows the late group, the threat is defined as a potential security disorder or violation to exploit the vulnerability of an IoT system or assets and can be intentional or unintentional, while the attack is defined as an intentional unauthorized act targeting IoT system or asset. A detailed contrast between these concepts is presented in Table 3.2.

Table 3.2 Comparison between threat and attacks in IoT environments

Features	Threat	Attack
Definition	Situation that might lead to damage to the IoT system	An intended action to reason damage to IoT system
Intentional	Threats can be deliberate (system failure) or unintended (e.g., natural disaster)	The attack is a deliberate action. An attacker has a motive and plans the attack accordingly
Initiation	Can be instigated by either foreigner or a relevant IoT entity	Usually instigated by a foreigner
Malicious	Can be malicious or not	The attack is always malicious
Degree of damage	The opportunity to damage or modify IoT information fluctuates from a low level to an extreme level	The opportunity to damage or modify IoT information alternation is very high
Detection	challenging to discover	Relatively simple to detect
Defense	It could be avoided by regulating the vulnerabilities	It could not be avoided by just regulating the vulnerabilities

3.2.2 IoT Vulnerabilities

The idea of IoT vulnerability is commonly defined as a point of fragility in the design, deployment, or functionality of the physical device, communication, and system. Vulnerabilities are omnipresent, and innumerable emerging vulnerabilities are revealed on a daily basis. Several web portals and online databases have been developed and make available to offer the community automatic and detailed updates concerning the recently detected IoT vulnerabilities.

Vulnerabilities might be shortcomings in software quality, device's physical security, system configuration, networking settings, or fitness of security protocol for the underlying ecosystem. This span everything in the device ranging from hardware design flaws to the operating system and the applications it hosts. Typically, attackers are conscious of the possible IoT vulnerabilities. They usually strive for discovering the IoT vulnerabilities that are simplest, least expensive, or swiftest to manipulate. They also aim to create a for-profit market of malevolent hacking founded on the dark web configurations; malicious hackers realize the notion of **return-on-investment** (**ROI**) in a good way. To sum up, the IoT threat can be viewed as the possibility for the exploit, while the vulnerability can be viewed as the goal of the genuine exploit that the threat actor seeks for [10, 46].

3.2.3 IoT Risks

Straightforwardly, the concept of risk in IoT ecosystems could be characterized as the entity's *exposure to loss*. The concept of Risk varies from vulnerability as it hinges on the possibility of a specific condition, incident, or attack and has a robust relationship to the attacker's motives. Also, it relies on how major is the effect of one or an entire set of attack occurrences. In contrast, the IoT vulnerability did not explicitly refer to effect or possibility but is the inherent weak point itself, which might be simple or difficult to exploit and thereby might lead to a minor or substantial loss as soon as being abused. A typical example of this, an operating system might have a significant vulnerability in its thread separation strategy enabling an unreliable thread to log on and modify the virtual memory of other processes. Such vulnerability might be exposed and undoubtedly epitomizes a weakness, but when the system is in an internet disconnected state, the vulnerability might represent slightly if any risk disclosure. Conversely, when the operating system is connected in some way to the Internet, the degree of risk might raise up as the attacker can discover a sensible way for introducing hostile shell code that exploits the thread separation vulnerability and lets the attacker presume proprietorship of the device.

The Risk could be handled via threat modeling, which facilitates discovering the subsequent factors: (1) Influence and total expenses of a compromise. (2) The degree to which the target is valuable with respect to attackers. (3) Expected talent and motives of the attackers. (4) A priori knowledge of a system's vulnerabilities such as those encountered throughout threat modeling, penetration check, community advisories, etc. Risk handling or management depends on the careful application for alleviating different categories of possible IoT vulnerabilities that can be targeted by the threat actors. Intuitively, some IoT vulnerabilities are not identified in advance, which is known as zero-days (0 days) vulnerabilities [47].

3.3 Today's IoT Attacks and Countermeasures

This section taxonomizes and discusses different contemporary IoT attacks according to IoT assets inferred from the layered architecture of the IoT system presented in Chap. 2. In particular, the IoT asset can be classified based on its threats and attacks options on IoT constituting layers into four types including physical devices, software, protocols, and data. Also, the discussion pinpoints a set of defense methods for securing IoT ecosystem [48].

3.3.1 Physical IoT Attacks (Hardware-Based)

Generally, IoT software is prone to a wide variety of cyber-attacks. since IoT integrate both the physical and cyber world together, the hardware elements of the physical side of IoT-based systems (i.e., sensors, controllers, RFID readers, etc.) are also susceptible to a variety of attacks, known as physical attacks. this category of attacks can be considered as steppingstone for different forms of IoT threats and attacks, for example, deactivating a building's warning system could lead to theft or other associated destruction. Likewise, a substitution of the device with a malevolent device could result in leakage of confidential data [48]. The Physical IoT attacks are generally categorized as either passive or active attacks, and the detailed description of common physical attacks, their targets, and possible countermeasures are presented in Table 3.3.

3.3.2 Software-Related IoT Attacks

In IoT environments, the data threats are not as dangerous as software threats security. In some incidents, even when the attackers already take the control of IoT application, they might still be unable to access to make use of the data if it was safely stored, or effectively encrypted, however, attacking the software can make them capable to perform other destructive data-irrelated activities including managing the IoT devices, interrupting system processes, or transmitting junk to other IoT elements [48]. In view of this, the software intended IoT attacks can be divided into three main categories namely Application-related IoT attacks, Operating system-related IoT attacks, and Firmware-related IoT attacks. an overview of different categories of software-related attacks, their description, compromised targets, and the related countermeasures is introduced in Table 3.4.

3.3.3 Data-Related IoT Attacks

This section argues the potential IoT attacks that mainly target IoT data in different locations of IoT systems including local data at IoT devices, remote and central stored data at cloud datacenters, data at edge nodes, or intermediate data at fog servers [48]. To this end, Table 3.5 presents a detailed description of common physical attacks, their targets, and possible countermeasures.

3.3.3.1 Protocol-Related IoT Attacks

Generally, IoT devices are equipped with a variety of connectivity practices partitioned roughly into two leading types, namely wired and wireless protocols. Again, the wired linking typically necessitates a physical medium (i.e., ethernet cables) among IoT devices, while wireless connection usually operates over radio waves. Each type of these connectivity technologies has a number of key features like topology, power consumption, data rate, scale, spectrum, etc. In view of this, the major IoT attacks aiming at the highly popular connectivity protocols, their description, security target, and possible countermeasures are tabulated in Table 3.6.

Table 3.3 Overview of hardware-related IoT attacks and corresponding countermeasures

Attack	Target	Descriptions	Countermeasures
Camouflage	All	The attackers physically injecting a fake object into a network of objects to pretend as a typical object to manipulate and resend the packets for malicious purposes [49]	Authentication, encryption, safeguarding firmware upgrade, hashing approaches [50]
Hardware Trojan	All	It is broadly known that the core threat in an amalgamated circuits is their exposure to physical Trojan attack, which seeks to maliciously adjust the integrated circuit to reach and access relevant sensitive data and firmware. This attack happens during the design phase and stays inactive till getting a prompt or an incident from its engineer[51]	Side-channel signal analysis, Trojan activation [51]
Malicious code injection	All	An adversary seeks to physically inject a malevolent code into an IoT object with the main aim to gain complete control of the IoT system [52, 53]	Static analysis, dynamic detection, firewalls [52]
Malicious node injection	All	The attackers seek to gain unauthorized access by inserting a malevolent object among legal ones in the IoT network, with the main aim to inject misleading data to obstruct messages transport, and possibly regulate the whole network. [46, 54]	Tamper proofing and self-destruction, intrusion detection
Object jamming	All	Despite the advantages of applying wireless networking to many IoT applications, its indicators might simply be hampered by means of a jammer [55, 40]	Precedence communications, territory mapping, distributed spectrum, low down duty cycle [56]

(continued)

Table 3.3 (continued)

Attack	Target	Descriptions	Countermeasures
Object replication	All	Attackers can attach an extra entity to the IoT network. i.e., a malevolent entity can be included by replicating the identity of entity. hence, this attack could reason an enormous decline in the communication performance. Also, it could corrupt or misdirect the received packets using a mischievous object, letting the attackers accessing confidential information and get the confidential keys [7, 57]	Computationally-efficient cartographic techniques, Hashing approaches [7, 58]
Object tampering	All	Among the physical IoT attacks, the object tampering seeks to extract cryptographic secrets to alter the embedded circuit, firmware, or operating system. A typical example of this attack is the use of malevolent thermostat to replace Nest thermostat [59, 60]	Self-annihilation, lowering data drip [59], entity integrated physical unclonable function, tamper proofing approaches [26]
Outage attacks	All	An attacker might stop a set of IoT objects from functioning in uncommitted environments owing to either power shutdown or utilizing much power	Secure physical design [27]
Physical damage	All	The unattended nature of IoT environments makes the IoT objects substantially vulnerable to physical attacks, the coolest one of them seek directly impair the physical modules [36]	Self-annihilation, protected physical construction, tamper-proofing [8]
RF interference	All	Transmitting a vast number of noisy signals across radio frequencies, with the main aim to interrupt radio communication i.e., RFID	Transmitter verification, cryptographic signatures, watermarking, intelligent authentication, anomaly detection [61, 62]

(continued)

Table 3.3 (continued)

Attack	Target	Descriptions	Countermeasures
Side-channel attack	S1, S4, S7	An attacker is aimed to break the IoT security mechanisms (i.e., encryption) through examining side-channel data radiated via IoT entities. This includes time, fault, electromagnetic, or energy analysis attacks [63]	Blockading, separation, destroy control, snooze control, tamper-proofing and self-destruction, channel attack detection, obscuring approaches [64]
Social engineering	All	An attacker manipulating physical IoT hardware to disclose precious and sensitive data could be viewed as a physical attack as the attackers might physically adjust the IoT devices to reach their targets [65]	Filtering tools, backup techniques, alerting and scanning software, biometric solutions, anti-social engineering framework, tamper-proofing, and self-destruction [66]
Tag cloning	S1, S3, S4, S7	Thanks to the attachment of tags on various thing, they exposed to be attacked physically. An attacker might effortlessly take over the tags and create a mockup of them resembling the genuine tags to infect or impair an RFID system through misleading the RFID readers [67]	Authentication methods, encryption approaches, hashing approaches [28], kill sleep command, segregation, blocking, space assessment. PUFs assimilation into RFID tags [68, 69]

3.4 IoT *Attack Surfaces*

This section mainly discusses surfaces of attack in IoT ecosystems and the possible threats associated with each surface. In view of this, the surfaces of IoT attack can be classified into four attack surfaces namely physical surface, network surface, cloud surface, application surface.

3.4.1 *Physical Surface*

Physical devices essential and indispensable constituents of the physical layer of IoT architectures, for instance, radio frequency Identification (RFID), which are a major element of IoT-based systems. Within an internet-connected embedded system, RFID performs a considerable responsibility in designing microprocessors for wireless

Table 3.4 Overview of software-related IoT attacks and corresponding countermeasures

Attack	Target	Descriptions	Countermeasures
Application-related IoT attacks			
Distributed denial of service (DDoS)	S1, S3, S8	An effort to exploit the vulnerabilities of the application layer of the IoT for interrupting the regular stream of traffic to a website, applications, or service to prevent delivering their content to the user. the botnet is a common way to establish DDoS [57, 70]	flow telemetry analysis, anomaly detection, access control lists [71]
The exploitation of a misconfiguration	All	Improper configuration of amenities (i.e., databases, hosting system) related to IoT application enables the attacker to easily access an IoT application [72]	Robust application design, execute checks and inspections constantly [73]
Malicious code injection	All	An invader inserts a malevolent code into various packets with the main aim to sneak or change confidential application data [74]	Static analysis, dynamic detection, firewalls [52]
Malware	All	The practice of contaminating web applications with malevolent software is termed malware. In Recent Times, a massive number of malwares has been devised for attacking IoT systems [75, 76]	Control flow side-channel assessment [77], software integrity validation, malware detector [78], Security updates [79]

(continued)

Table 3.4 (continued)

Attack	Target	Descriptions	Countermeasures
Path-based DoS attack	S1, S2, S3, S7	An attacker replays some packets to the network or injects malicious code into the packets. It could exhaust the IoT network by transmitting a massive amount of legal packets preventing other applications or services from transmitting messages to the support [80]	Mixing anti-replay security and packet authentication [81]
Reprogram attack	S1, S2, S4, S7, S11	Remote reprogramming of IoT applications is achievable via a network programming system. After the reprogramming procedure became not secure, the attacker might hijack this process to regulate a huge portion of the IoT network [82]	Encryption, update, secure the reprogramming process [41, 83]
Operating system-intended IoT attacks			
Backdoors	All	Developers have been designed various IoT operating systems such as RTOS and Contik, which might include a backdoor to be reprogramed by attackers to gain control of the system anytime [84]	offline inspection, blind backdoor exclusion, online examination, and upright backdoor exclusion [84]
Brute-force search attack	All	It refers to an attempt to attack IoT ecosystem by breaching the relevant security procedures like authentication and cryptography utilizing automated trial-and-error methods [85]	Securing firmware update, cryptography methods, deception model [86]

(continued)

Table 3.4 (continued)

Attack	Target	Descriptions	Countermeasures
Phishing attack	All	An attacker seeks to reach the passwords, credit cards, bank account, and other confidential system data by hacking social accounts, chats, email, phones	Cryptographic methods
Unknown attack	/	It worth mentioning that some familiar exposures and disclosures files have not supplied sufficient data to characterize the attack's prerequisites, which categorize as an anonymous attack [87]	AI intrusion detector, fingerprinting [88]
Virus, worm attack	All	These days, several worms and viruses, such as Mirai, Stuxnet, and Bricker Bot, were developed to hack some system flaws like the absence of update processes noticed in IoT devices and nodes [89]	Flow monitoring, security upgrades, software integrity validation, side-channel assessment defensive Software [90]
Firmware-related IoT attacks			
Control hijacking	All	The process of changing the normal flow control of the IoT object firmware by injecting a malicious code is known as a control hijacking attack [91]	Runtime code injection, reliable programming, inspection software, screening techniques [92]
Reverse Engineering	All	An attempt to analyze the firmware of IoT devices to reach sensitive data i.e., users' credentials [93]	Self-destruction and Tamper proofing, IC/IP Obfuscation and encryption [94, 95]

(continued)

Table 3.4 (continued)

Attack	Target	Descriptions	Countermeasures
Eavesdropping	S4, S7	A passive attack that monitors the weakness in the transmission of packets between IoT nodes during the process of firmware update, to obtain sensitive data. It also might retransfer the packets to create a restate attack [96]	Thorough inspection, firewalls, a secure design [97]
Malware	All	Attempt to change the behavior of the IoT system by contaminating its firmware with a mischievous code. Numerous malware has been noticed in the i.e., BASHLITE, Hydra, and Darlloz [98]	Flow monitoring, security upgrades, software integrity validation, side-channel assessment defensive Software, malware detector [78]

Table 3.5 Overview of data related IoT attacks and corresponding countermeasures

Attack	Target	Descriptions	Countermeasures
Account hijacking attacks	All	An attacker can exploit social engineering and weak passwords to carry out an account hijacking to compromise, control, and send sensitive data. In cloud-assisted IoT, various on-request services are provided using different application program interfaces (API) that suffer from unsatisfactory authorization and data validation [99]	Identity and access management, authentication, multi-factor authentication guidance, dynamic credentials [100]
Brute-force attacks	S4, S6	An active attack hinges on a trial-and-error strategy to obtain some data such as passwords, finance, identifiers. It employs automatic software to engender a massive amount of successive suppositions to decrypt the ciphertext [85, 86]	Locking out IP address, discovery tools, brute force site scanning tools [101]
Data exposure attacks	S4, S6	The possibility that malicious objects easily access some insecure data sources because of the unavailability of encryption and other security management [102]. The distributed storage of IoT data across multiple geographical countries might enable attackers to access the data at some insecure holders [103]	Security management, robust encryption algorithms, strict access controls [53, 104]

(continued)

Table 3.5 (continued)

Attack	Target	Descriptions	Countermeasures
Data leakage Attacks	All	This refers to attacks that seek for deliberate or accidental release of protected or private/confidential data either edge, fog, cloud, or on IoT devices [105]	Fragmentation-redundancy-scattering (FRS) [106], encryption, digital signature [107]
Data loss Attacks	All	An attacker intends to make IoT devices, fog/edge nodes, and cloud providers completely or partially losing their hosted data, triggering damaging penalties. i.e., a ransomware attack	Robust key generation, secure storage, and management [108], backup and preservation policies
Data manipulation attacks	All	Unlawful manipulation of IoT data could be accomplished in two ways. First, taking advantage of various weaknesses in API as including cross-site scripting and SQL injection attacks. second, exploiting fragile security procedures like weak authentication [109]	File integrity monitoring, logging activity, web application scanners [109]
Data scavenging attacks	S4, S6, S10	The recoverable characteristic of IoT data renders it vulnerable to many attacks when it was not accurately obliterated or deleted [110]	Symmetric key cryptography [111]
Denial-of-services	S1, S3, S8	The kind of attack that seeks to make IoT data unavailable or authorized users or software. It exploits the vulnerabilities of the system APIs [41, 70, 80]	Policies provided by providers [111]

(continued)

Table 3.5 (continued)

Attack	Target	Descriptions	Countermeasures
Hash collision	S4, S6	It is an effort for finding out dual input sequences of a hashing operation that provides one hash value. Since hashing tasks come up with the input of varying lengths and a small, static length production, the likelihood that couple of distinct inputs create the identical outputs referred to as a collision [112]	Contemporary hashing systems such as $SHA-2, SHA-3$ [113]
Insecure VM migration	All	Adversary might gain illegal access to IoT throughout the immigration operation of a virtual machine to a malevolent or a dependable host, which might reveal its data to the public [114]	Defense guidance for inhabit resettlement of virtual machine, VNSS presents security via the live migration of virtual machine [115]
Malicious VM creation	All	As numerous virtual machine images are installed in unattended IoT ecosystems, an attacker might create a legitimate virtual machine account comprising a malevolent code such as a Trojan horse [116]	Mirage [111]
Virtual machine (VM) escape	All	It is an attempt to make the use of vulnerabilities of hyper-visor with the main aim to dictate the fundamental infrastructure greedy for its design elasticity and code intricacy, which agree with enterprises requirements [117]	Trustworthy virtual datacenter, dependable cloud computing platform, hyper safe, appropriately designing the host/guest interaction [25]

(continued)

Table 3.5 (continued)

Attack	Target	Descriptions	Countermeasures
VM hopping	All	It is an attempt to exploits the hypervisor's weakness that tolerates a virtual machine to gain access to another one for data theft and other malicious purposes [118, 119]	Access control [118], virtual machine debugging tools, high assurance platform (HAP) [120]

communication [142]. RFID is always tagged to any IoT objects including humans, physical objects, and animals. The main attribute of RFID tags is automated recognition beyond a distinctive recognizer that entails speedy data communication between the readers and tags [143]. The principal purpose of RFID technology is to oversee the procedure of physical in the meantime bridging the interaction between real and computer-generated environments. Thus, these small physical things could be consumed in an extraordinarily broad scale of IoT applications [142]. Nevertheless, the majority of physical devices undergo several security-associated concerns. One more element of the physical device surface is the sensor node, which primarily comprises sensors employed for sensing the environments and actuators employed for actuating devices compliant with a particular set of predefined directives. Besides, sensor nodes are generally known to have limited resources and always exhibit elevated latency. The majority of the physical IoT devices having limited computational resources and usually host important information, making them a probable surface for either physical or cyber attackers; for instance, the attackers/intruders might exploit the physical devices to track their locations, overflow them with countless entree requests that reason DoS, spoofing, counterfeiting, or other forms of attacks [144, 145]. Furthermore, the physical surface is extremely exposed to physical threats for the reason that it is the most physically reachable surface for attackers and intruders.

3.4.2 Network Surface

IoT environment usually encompasses internet-connected physical things over wired and wireless networking tools. Sensor networks are considerable constituent assets for an IoT-based system although they could be created with no IoT-based system. An IoT-based system generally involves Sensor networks along with wired and wireless communication, therefore establishing a larger-scale and continuously expanding network surface, which typically has an increased number of potential vulnerabilities. Additionally, IoT-based systems are subject to a variety of novel security threats derived from the inherent characteristics of either wired or radio sensor

Table 3.6 Overview of protocol related IoT attacks and corresponding countermeasures

Attack	Target	Descriptions	Countermeasures
Radio-frequency identification (RFID)			
Replay	S3, S4, S6, S7	An attacker might exploit tags' replies to phony readers' encounters. Where the transferred information from tags to readers is caught, recorded, and echoed afterward to the receiver, resultant in forging the availability of the tag [102]	Time-based approaches, counter-based approaches, challenge-response method [108]
Spoofing	All	It denote an attack occurring once a malevolent tag imitates as a lawful tag and acquires unlicensed access. It is usually employed to eavesdrop on the data emanating from the legitimate tag and transmitting its detained data to the fake ones [26]	RFID authentication and encryption techniques [111]
Tracking	S4, S7	It is a form of direct attack targeting a system user or a victim. Recently, corporations may possibly put RFID tags on several domestic objects. Tracking goods utilizing RFID tags might jeopardize the confidentiality of humans by tracing their actions and create an accurate report about their purchasing [121]	Sleep control, Kill control, remoteness, mysterious tag, impeding [57]
Unauthorized access	All	Thanks to the absence of authentication in the RFID system, tags are likely to undergo an unauthorized attack with the main aim to access and manipulate relevant confidential data [122]	Network authentication [122]

(continued)

Table 3.6 (continued)

Attack	Target	Descriptions	Countermeasures
Virus	All	An RFID can be viewed as an inadequate ecosystem for tag-alike viruses having a tiny storage resource. Nevertheless, RFID tags can be employed as a channel to propagate the infected software [108]	Well-formed middleware for impeding weird bits from the tag, limits inspection, parameters check [108]
Eavesdropping	S4, S6, S7	For RFID ecosystem, wireless transmission from tags to readers is prone to have eavesdropped. Generally, eavesdropping gets triggered when an adversary obtains the non-encrypted data communicated between tag and reader because of low memory capacity [97]	Backend for data shifting methods, encryption approaches
Man in the middle (MITM)	S4, S6, S7	Adversary might capture and change the channel of communication from RFID readers to tags. It can be deemed as a genuine time attack, showing and adjusting the data prior to being received by a legitimate entity [102]	Encryption of the RFID communication channel [123], authentication techniques
Killing Tag	All	It is an attempt to block tag's interaction with the corresponding reader by making tags impractical to be read, and hence, it is indispensable to ensure that RFID tags are not destroyed with an unlawful parity [124]	Users or objects authentication [125]

Near-Field Communication (NFC)

(continued)

Table 3.6 (continued)

Attack	Target	Descriptions	Countermeasures
Eavesdropping	S4, S6, S7	In an NFC system, data exchange between two devices happens in the close vicinity making the communication channel between them vulnerable to an eavesdropping attack. An attacker intercepts this channel utilizing a robust antenna or be in the vicinity of the transmission scale	Secure channel (authentication and encryption) [44]
Relay	S3, S4, S6, S7	Attackers redirect the calls from the readers of object to a malevolent one and replay backward its reply rapidly [126]. It severely relies on the implementation of the application protocol data unit instructions (ISO/IEC1443)	Timing [127], distance bounding of cryptographic challenge-response couples [128]
Man in the middle (MITM)	S4, S6, S7	Adversary can capture the data, alter and send it to malevolent things in close vicinity making such attacks very complicated, encryption methods also make it difficult to succeed if they are fulfilled correctly [129]	A secure channel between the NFC objects
Data corruption	All	An attacker possesses the ability to disrupt communication channels between NFC-armed IoT devices by altering the transmitted data to be unreadable leading to a denial of services attack [130]	The discovery of RF spheres throughout the communication of data [44]
Data modification	All	An attacker possesses the ability to attacker modify the content of communicated data between NFC-armed IoT devices [131]	Channel securing, Baud rate adjustment, constant checking of RF arena [44]

(continued)

Table 3.6 (continued)

Attack	Target	Descriptions	Countermeasures
Data insertion	S1, S2, S7, S8	An attacker attempt to inject some information into transmitted data when the NFC-armed device needs a long time to reply [130]	Immediate entities response, securing the channel between two entities [130]
Bluetooth			
Bluesnarfing	All	An attacker seeks to gain unlawful access to Bluetooth devices with the aim to capture their information and forward the incoming requests to another device [49]	Setting mobiles on non-ascertainable style [49], keep on disconnected [64], validate next transmission
BlueBugging	All	An adversary might use some weaknesses in legacy firmware to get into the victim's device to eavesdrop on phone calls, messages, emails, and link up to the internet without the awareness of the owner [49]	Updating software and firmware, apply signatures to RF signals [132]
Bluejacking	S1, S2, S7, S10	An attacker can exploit the ability to transmit a radio business card to send an assault card; nonetheless, this necessitates the attacker to be very close i.e., within 10 m from the victim's device [49]	Non-ascertainable style [49], keep on disconnected [64], validate next transmission [133]
DoS	S1, S8, S3, S9, S7	Repetitive usage of Bluetooth device to transmit a pairing demand to the devices of targets to enable an attacker to trigger DoS attacks. This can be considered as one of the easiest ways for draining a device's battery [50]	Retaining a directory of untrustworthy entities [65]

(continued)

Table 3.6 (continued)

Attack	Target	Descriptions	Countermeasures
Interception	S4, S7, S10	Bluetooth interference does not necessitate complex or costly hardware. Instead, many affordable hardware is enough for this i.e., Ubertooth [134]	Encryption, use of long PIN codes [132], private pairing venues [133]
Hijacking	All	An unauthorized third party is employed to regulate the beacon configurations. spoofing and DoS might occur as a result [135]	Cloud-founded token certification, protected communication channels, applying lock-out to the software [132]
Spoofing	S6, S7, S1	Since the beacon is openly transmitted, an attacker can use a sniffing tool to capture the beacon's A universally unique identifier (UUID), mimic the beacon and break the application rules, and perform illegal access to the services [135]	Protected rearranging, secure UUID, arbitrarily spinning UUID [66]
Wifi			
(Fluhere, Shamir, and Mantin) FMS attack	S1, S4, S7, S6	The attackers exploited the vulnerabilities of the wired equivalent privacy (WEP) protocol to compromise it. It is a torrent cipher attack that seek to retrieve the keys employed to encrypt the communication by realizing initial vectors. Nevertheless, the probability of an attack on RC4-based SSL(TLS) is extremely complicated because of the reliance of key generation operation on hashing function [136]	Applying RC4-based SSL (TLS), the usage of improved security procedures like IPsec [67]

(continued)

Table 3.6 (continued)

Attack	Target	Descriptions	Countermeasures
Korek attack	S4, S7, S10, S8, S11	A new attack targeting the WEP protocol [137], which depends on FMS attack to discover the key. But it employs an A-neg attack to lessen the probabilities of key generation to enable discovering the key rapidly [55]	The usage of an extremely small rekeying time, deactivating the transmitting of MIC breakdown report, incapacitating TKIP and utilizing a counter mode CBC-MAC (CCMP) only network [138], the usage of advanced-level security procedures like IPsec, DTLS, HTTP/TLS or CoAP/DTLS, DTLS for constrained application protocol [139]
Chopchop attack		Rather than targeting the weakness of the RC4 algorithm, this attack emphasis on the design flaws of in WEP protocol i.e., CRC 32 checksum and lack of replay protection. Chopchop attacks enable the attacker to decrypt the communication messages without realizing the key [137]	
PTW attack		By 2007, the Pyshkin Tews Weinmann (PTW) attack has presented two principles: (1) Jenkins's connection suggested to speculate the key with least tries, and (2) numerous bytes estimate rather than speculating bytes independently [140]	
Dictionary attack	All	A method wherein an attacker might break into a password-guarded Wi-Fi by speculating its phrases by trying billions of opportunities, like as a dictionary of words [137]	The use of salt technique [141]

networks. These latest threats are launched as soon as standard sensor networks are promptly incorporated into IoT networks. The straight incorporation of a wired or wireless sensor network into an IoT network imposes many problems owing to the fact that standard sensor networks become no longer reliable within IoT conditions. For instance, the resistance of wireless sensor networks makes this network wholly susceptible to attacks/intrusions under IoT environments [57, 146]. On the other hand, the invaders can launch a kind of threat that targets the transmission protocols with the main aim to cause network failure. Therefore, developing a reliable communication protocol is essential for the security of IoT networks [144, 147]. Besides, the attacks might be launched at specific communication endpoints (i.e., port) by exploring and checking accessible ports. Exposure of accessible ports can motivate the attackers/intruders to initiate an incident on the IoT applications or services functioning on those accessible ports. This form of attack could acquire thorough information about the network, including media access control (MAC) address, internet protocol (IP) address, routers, and gateway settings [148, 149].

With the proliferation of IoT, network connectivity, agility, and cooperation among different entities is witnessing great expansion. This nature further expands the surface of network services resulting in many of security vulnerabilities, threats, and risks including interruption, hacking, MITM attack protocol tunneling, admission spoofing, Denial of service, and interception [150]. Moreover, the Internet network, which is a major constituent for connecting IoT devices, has diverse actors oscillating from a local network area (LAN) to a worldwide network area (WAN) and from business contributors to human contributors, in that way linking an extensive scope of edges, nodes, and servers [101]. Firstly, Internet connectivity enable delivering a broad-scale of applications and services that may operate using the aggregated sensory information to realize a completely functional IoT system for delivering smart applications and services. At the same time, the constant utilization of conventional Internet protocol (i.e., TCP/IP, UDP/IP, etc.) to interconnect millions or even billions of physical things and devices globally is extremely exposed to a wider range of IoT threats endangering the security and privacy of overall IoT environments [151].

3.4.3 Cloud Attack Surface

Cloud computing offers a group of pioneering applications that are launched to provide admission to save and practices for acquiring data from wherever and whenever; therefore, the necessity for hardware resources on the client-side is either constrained or abolished [152]. Cloud computing offers a powerful computing paradigm that can be exploited as a foundation technology to build improved and pioneering IoT applications by facilitating remote access to pooled resources upon request. Cloud computing has considerable features that greatly bring many advantages to IoT-based systems, such as shifting applications and services from client-side to Internet side, optimized provision of computational and storage resources, and efficient energy management [153]. The incorporation of cloud into IoT-based systems (i.e., cloud-assisted IoT) poses huge openings for IoT ecosystems by benefiting from the plentiful cloud resources in such a way that defeats the resource constraints which is a major limitation of IoT, e.g., computing resources and power facilities [154]. The incorporation of the cloud into IoT also comes up with distinct occasions for the cloud. The cloud could exploit physical devices to act as a channel to be assimilated into realistic applications via active and distributed manners, therefore, delivering cloud services to a broad scale of users [153]. Nevertheless, this cloud-assisted IoT-based systems generally result in multiple security threats as the distributed nature of this system render it susceptible to a different set of IoT attacks. For example, malicious attacks that could take advantage of weaknesses in cloud system to acquire unapproved entrance to data, services, or applications (e.g., cross-site scripting (XSS), SQL injection flaws, cross-site request forgery (CSRF), and insecure storage [155]). Unsatisfactory data integrity mechanisms could enable the attackers/intruders to avoid the authorization procedure to promptly access the database [156]. Furthermore, the vulnerabilities of cloud virtualization software could be abused by adversaries to gain illegal access, hack different virtual machines, or change different system privileges, etc. [112, 113].

Cloud computing exhibits significant effects on the privacy and security of information in IoT environments. Confidentiality and privacy risks vary substantially in accordance with the conditions and circumstances related to the clients and the cloud service providers. Nevertheless, the interaction between IoT devices and cloud data centers come up with a number of privacy interests like revealing extremely secret information (e.g. credentials, medical records, in-home sensory data, etc.). Privacy can be considered as a crucial trust aspect that motivates users to stop using IoT devices or services. Consequently, development ought to be complemented with valuable privacy preservation for a thriving IoT deployment [157–159]. Besides, the multi-tenancy nature of cloud computing environments enables manifold clients to store up their own data on remote cloud server thru the application program interface (API), which makes it vulnerable to some IoT attacks that might result in the leakage of private information. In this context, the users' data was stored at a single locality, which could be accessed by any of these users. Attackers can exploit the gaps in the API or inject some code into the cloud system to gain access or perform an unauthorized operation on users' data [112]. Also, permitted cloud users may possibly abuse their allowable access to obtain unlawful privileges and initiate attacks, i.e., interior DoS [41].

3.4.4 Application Surface

The majority of IoT applications or services provided are delivered to the users remotely using a mobile or Web interface. For instance, in a smart buildings, the smart appliances are wirelessly linked to users' smart devices (smartphone, or tablet), which are employed to control the home appliances mostly using some mobile applications or web interfaces sometimes [160]. Also, mobile applications have been widely developed for different IoT targets, smart vehicles, lights, parking, belts, shoes, sunglasses, and so many others that are making IoT-based devices to be under the remote control of mobile applications. With the advent of IoT technologies, the real and virtual sides of the worlds become deeply combined, and almost immediately the distinction between the two sides will come to be indescribable. IoT devices can communicate with each other immediately. This setup could be completely realized with the support of mobile applications [153]. Mobile phones have come to be omnipresent for the reason that the widespread services they deliver to clients throughout their hosted applications, which make them dominate the marketspace owing to their open design and the reputation of their APIs amongst developers [161, 162]. Nevertheless, the open nature of mobile OS allows the IoT users to download a variety of applications including mischievous applications developed with third-party without comprehensive security inspections [163]. The expanding acceptance of mobile devices has been attracting the attention of malware developers, shadowed by massive growth in malware attacks i.e., Android malware [164]. Malware attackers could control smartphones by exploiting system susceptibilities, getting personal user credentials, or creating botnets. Additionally, Android applications might announce confidential information thoughtlessly or malevolently. Therefore, their operational behaviors, functioning models, and practice patterns ought to be identified to create an intelligent security solution to handle the security issues of mobile devices [165]. A typical example of mobile application attacks includes eavesdropping, tracking, bluejacking, bluesnarfing, and DoS [166].

3.5 Summary and Learnt Lessons

This chapter provides an in-depth discussion about the IoT security requirements, IoT threats, IoT vulnerabilities, Risks, and attacks providing a taxonomy for categorizing different threats, indicating the main IoT attack surfaces. In the nutshell, this chapter delivers a useful insight about different aspects of IoT security that could be pointed out as follow:

- IoT environments are complicated and encompass multidisciplinary cyber-physical components. Hence, traditional information assurance is not enough to cope with this nature. Thus, the concept of information assurance is extended to define the security requirements of any IoT system. Fulfilling these requirements is challenging owing to the expanding scale of attack surfaces, thereby developing security solution have to encompass holistic considerations.
- This chapter provides a detailed discussion about the IoT threats, vulnerabilities, and risks and how these concepts differ from each other. A Taxonomy of categorizing IoT threats is also presented.
- To keep up with today's IoT attacks, the discussion also encounters the prevailing IoT attacks and the possible countermeasures. Later, the discussion deliberates the potential attack surfaces and the related challenges.

References

1. M. Nieles, K. Dempsey, V.Y. Pillitteri, NIST Special Publication 800-12 Revision 1—An introduction to information security, NIST Spec. Publ. (2017)
2. M. Nieles, K. Dempsey, V.Y. Pillitteri, NIST SP800-12 Revision 1: An Introduction to Information Security, NIST Spec. Publ. (2017)
3. B. Russell, D. Van Duren, Practical Internet Of Things Security (2016)
4. I. Makhdoom, M. Abolhasan, J. Lipman, R.P. Liu, W. Ni, Anatomy of threats to the Internet of Things. IEEE Commun. Surv. Tutorials (2019). https://doi.org/10.1109/COMST.2018.287 4978
5. A. Aldweesh, A. Derhab, A.Z. Emam, Deep learning approaches for anomaly-based intrusion detection systems: a survey, taxonomy, and open issues. Knowl.-Based Syst. (2020). https://doi.org/10.1016/j.knosys.2019.105124
6. N. Chaabouni, M. Mosbah, A. Zemmari, C. Sauvignac, P. Faruki, Network intrusion detection for IoT security based on learning techniques. IEEE Commun. Surv. Tutorials. (2019). https://doi.org/10.1109/COMST.2019.2896380
7. H.I. Ahmed, A.A. Nasr, S. Abdel-Mageid, H.K. Aslan, A survey of IoT security threats and defenses. Int. J. Adv. Comput. Res. (2019). https://doi.org/10.19101/ijacr.2019.940116
8. N. Agrawal, S. Tapaswi, Defense mechanisms against DDoS attacks in a cloud computing environment: state-of-the-art and research challenges. IEEE Commun. Surv. Tutorials (2019). https://doi.org/10.1109/COMST.2019.2934468
9. M. El-Hajj, A. Fadlallah, M. Chamoun, A. Serhrouchni, A survey of internet of things (IoT) authentication schemes. Sensors (Switzerland) (2019). https://doi.org/10.3390/s19051141
10. M.A. Al-Garadi, A. Mohamed, A.K. Al-Ali, X. Du, I. Ali, M. Guizani, A survey of machine and deep learning methods for Internet of Things (IoT) Security. IEEE Commun. Surv. Tutorials (2020). https://doi.org/10.1109/COMST.2020.2988293
11. S.V. Sudarsan, O. Schelén, U. Bodin, Survey on delegated and self-contained authorization techniques in CPS and IoT. IEEE Access (2021)
12. N. Neshenko, E. Bou-Harb, J. Crichigno, G. Kaddoum, N. Ghani, Demystifying IoT security: an exhaustive survey on IoT vulnerabilities and a first empirical look on Internet-scale IoT exploitations. IEEE Commun. Surv. Tutorials (2019). https://doi.org/10.1109/COMST.2019. 2910750
13. J.M. Hamamreh, H.M. Furqan, H. Arslan, Classifications and applications of physical layer security techniques for confidentiality: a comprehensive survey. IEEE Commun. Surv. Tutorials (2019). https://doi.org/10.1109/COMST.2018.2878035
14. C. Lesjak, T. Ruprechter, H. Bock, J. Haid, E. Brenner, ESTADO—enabling smart services for industrial equipment through a secured, transparent and ad-hoc data transmission online, in 2014 9th International Conference for Internet Technology and Secured Transactions ICITST 2014 (2014). https://doi.org/10.1109/ICITST.2014.7038800
15. T. Pereira, L. Barreto, A. Amaral, Network and information security challenges within Industry 4.0 paradigm, Procedia Manuf. (2017). https://doi.org/10.1016/j.promfg.2017.09.047
16. P. Autenrieth, C. Lorcher, C. Pfeiffer, T. Winkens, L. Martin, Current significance of IT-infrastructure enabling Industry 4.0 in large companies, in 2018 IEEE International Conference on Engineering, Technology and Innovation ICE/ITMC 2018—Proceedings, 2018. https://doi.org/10.1109/ICE.2018.8436244
17. N. Jazdi, Cyber physical systems in the context of Industry 4.0, in Proc. 2014 IEEE International Conference on Automation, Quality and Testing, Robotics AQTR 2014 (2014). https://doi.org/10.1109/AQTR.2014.6857843
18. J. Moyne, S. Mashiro, D. Gross Determining a security roadmap for the microelectronics industry, in 2018 29th Annual SEMI Advanced Semiconductor Manufacturing Conference ASMC 2018 (2018). https://doi.org/10.1109/ASMC.2018.8373213
19. N. Benias, A.P. Markopoulos, A review on the readiness level and cyber-security challenges in Industry 4.0, in South-East Europe Design Automation, Computer Engineering, Computer Networks and Social Media Conference SEEDA-CECNSM 2017 (2017). https://doi.org/10.23919/SEEDA-CECNSM.2017.8088234

20. A. Hassanzadeh, S. Modi, S. Mulchandani, Towards effective security control assignment in the Industrial Internet of Things, in IEEE World Forum Internet Things, WF-IoT 2015—Proceedings, 2015. https://doi.org/10.1109/WF-IoT.2015.7389155

21. C. Esposito, A. Castiglione, B. Martini, K.K.R. Choo, Cloud manufacturing: security, privacy, and forensic concerns. IEEE Cloud Comput. (2016). https://doi.org/10.1109/MCC.2016.79

22. P. Li, J. Su, X. Wang, ITLS: lightweight transport-layer security protocol for IoT with minimal latency and perfect forward secrecy. IEEE Internet Things J. (2020). https://doi.org/10.1109/JIOT.2020.2988126

23. S. Cavalieri, A proposal to improve interoperability in the industry 4.0 based on the open platform communications unified architecture standard. Computers (2021). https://doi.org/10.3390/computers10060070

24. K. Tange, M. De Donno, X. Fafoutis, N. Dragoni, A systematic survey of Industrial Internet of Things security: requirements and fog computing opportunities. IEEE Commun. Surv. Tutorials (2020). https://doi.org/10.1109/COMST.2020.3011208

25. S. Mahipal, V. Ceronmani Sharmila, Virtual machine security problems and countermeasures for improving quality of service in cloud computing, in Proceedings—International Conference on Artificial Intelligence and Smart Systems ICAIS 2021 (2021). https://doi.org/10.1109/ICAIS50930.2021.9395922

26. A. Juels, RFID security and privacy: a research survey. IEEE J. Sel. Areas Commun. (2006). https://doi.org/10.1109/JSAC.2005.861395

27. V. Sharma, I. You, K. Andersson, F. Palmieri, M.H. Rehmani, J. Lim, Security, privacy and trust for smart mobile-Internet of Things (M-IoT): a survey. IEEE Access (2020). https://doi.org/10.1109/ACCESS.2020.3022661

28. W.S. Blackmer, EU general data protection regulation (GDPR). Off. J. Eur. Union. (2016)

29. A. Alwarafy, K.A. Al-Thelaya, M. Abdallah, J. Schneider, M. Hamdi, A survey on security and privacy issues in edge-computing-assisted Internet of Things. IEEE Internet Things J. (2021). https://doi.org/10.1109/JIOT.2020.3015432

30. S.A. Hamad, Q.Z. Sheng, W.E. Zhang, S. Nepal, Realizing an Internet of Secure Things: a survey on issues and enabling technologies. IEEE Commun. Surv. Tutorials (2020). https://doi.org/10.1109/COMST.2020.2976075

31. Z. Lv, Y. Han, A.K. Singh, G. Manogaran, H. Lv, Trustworthiness in Industrial IoT systems based on artificial intelligence. IEEE Trans. Ind. Inf. (2021). https://doi.org/10.1109/TII.2020.2994747

32. Y. Xu, J. Ren, G. Wang, C. Zhang, J. Yang, Y. Zhang, A blockchain-based nonrepudiation network computing service scheme for industrial iot. IEEE Trans. Ind. Inf. (2019). https://doi.org/10.1109/TII.2019.2897133

33. S. Yang, F. Li, S. Trajanovski, R. Yahyapour, X. Fu, Recent Advances of Resource Allocation in Network Function Virtualization. IEEE Trans. Parallel Distrib. Syst. (2021). https://doi.org/10.1109/TPDS.2020.3017001

34. M. Serror, S. Hack, M. Henze, M. Schuba, K. Wehrle, Challenges and opportunities in securing the Industrial Internet of Things. IEEE Trans. Ind. Inf. (2021). https://doi.org/10.1109/TII.2020.3023507

35. iiconsortium, Industrial Internet of Things volume G4: security framework. Ind. Internet Consort. (2016)

36. E. Sisinni, A. Saifullah, S. Han, U. Jennehag, M. Gidlund, Industrial internet of things: challenges, opportunities, and directions. IEEE Trans. Ind. Inf. (2018). https://doi.org/10.1109/TII.2018.2852491

37. A. Laszka, W. Abbas, Y. Vorobeychik, X. Koutsoukos, Synergistic security for the Industrial Internet of Things: integrating redundancy, diversity, and hardening, in Proceedings—2018 IEEE International Conference on Industrial Internet, ICII 2018 (2018). https://doi.org/10.1109/ICII.2018.00025

38. V. Hassija, V. Chamola, V. Gupta, S. Jain, N. Guizani, A survey on supply chain security: application areas, security threats, and solution architectures. IEEE Internet Things J. (2021). https://doi.org/10.1109/JIOT.2020.3025775

39. F. Hussain, R. Hussain, S.A. Hassan, E. Hossain, Machine learning in IoT security: current solutions and future challenges. IEEE Commun. Surv. Tutorials (2020). https://doi.org/10.1109/COMST.2020.2986444

40. Y. Dong, P. Zhou, Jamming attacks against control systems: a survey. Commun. Comput. Inf. Sci. (2017). https://doi.org/10.1007/978-981-10-6373-2_57

41. M. Malik, Y. Singh, A review: DoS and DDoS attacks. Int. J. Comput. Sci. Mob. Comput. (2015)

42. F. Meneghello, M. Calore, D. Zucchetto, M. Polese, A. Zanella, IoT: Internet of threats? A survey of practical security vulnerabilities in real IoT devices. IEEE Internet Things J. (2019). https://doi.org/10.1109/JIOT.2019.2935189

43. F.O. Olowononi, D.B. Rawat, C. Liu, Resilient machine learning for networked cyber physical systems: a survey for machine learning security to securing machine learning for CPS. IEEE Commun. Surv. Tutorials (2021). https://doi.org/10.1109/COMST.2020.3036778

44. N. Wang, P. Wang, A. Alipour-Fanid, L. Jiao, K. Zeng, Physical-layer security of 5G wireless networks for IoT: challenges and opportunities. IEEE Internet Things J. (2019). https://doi.org/10.1109/JIOT.2019.2927379

45. D. Wang, B. Bai, W. Zhao, Z. Han, A survey of optimization approaches for wireless physical layer security. IEEE Commun. Surv. Tutorials (2019). https://doi.org/10.1109/COMST.2018.2883144

46. J. Liu, S. Zhang, W. Sun, Y. Shi, In-vehicle network attacks and countermeasures: challenges and future directions. IEEE Netw. (2017). https://doi.org/10.1109/MNET.2017.1600257

47. J.H. Kim, A survey of IoT security: risks, requirements, trends, and key technologies. J. Ind. Integr. Manag. (2017). https://doi.org/10.1142/s2424862217500087

48. H.A. Abdul-Ghani, D. Konstantas, M. Mahyoub, A comprehensive IoT attacks survey based on a building-blocked reference model. Int. J. Adv. Comput. Sci. Appl. (2018). https://doi.org/10.14569/IJACSA.2018.090349

49. L. Huang, C. Gao, Y. Zhou, C. Xie, A. Yuille, C. Zou, N. Liu, Universal physical camouflage attacks on object detectors, in Proc. IEEE Computer Society Conference on Computer Vision and Pattern Recognition, 2020. https://doi.org/10.1109/CVPR42600.2020.00080

50. M. El Massad, S. Garg, M. V. Tripunitara, The SAT attack on IC Camouflaging: impact and potential countermeasures. IEEE Trans. Comput. Des. Integr. Circuits Syst. (2020). https://doi.org/10.1109/TCAD.2019.2926478

51. K.G. Liakos, G.K. Georgakilas, S. Moustakidis, N. Sklavos, F.C. Plessas, Conventional and machine learning approaches as countermeasures against hardware Trojan attacks. Microprocess. Microsyst. (2020). https://doi.org/10.1016/j.micpro.2020.103295

52. D. Mitropoulos, D. Spinellis, Fatal injection: a survey of modern code injection attack countermeasures. PeerJ Comput. Sci. (2017). https://doi.org/10.7717/peerj-cs.136

53. J. Deogirikar, A. Vidhate, Security attacks in IoT: a survey, in Proceedings International Conference on IoT in Social, Mobile, Analytics and Cloud, I-SMAC 2017 (2017). https://doi.org/10.1109/I-SMAC.2017.8058363

54. Y. Xiao, Security in distributed, grid, mobile, and pervasive computing (2007). https://doi.org/10.1201/9780849379253

55. Y. Deng, T. Zhang, G. Lou, X. Zheng, J. Jin, Q.L. Han, Deep learning-based autonomous driving systems: a survey of attacks and defenses. IEEE Trans. Ind. Inf. (2021). https://doi.org/10.1109/TII.2021.3071405

56. A. Ahmed, U. Ashraf, F. Tunio, K. Abu Bakar, M.S. Al-Zahrani, Stealth jamming attack in WSNs: effects and countermeasure. IEEE Sens. J. (2018). https://doi.org/10.1109/JSEN.2018.2852358

57. J. Sen, A survey on wireless sensor network security. Int. J. Commun. Networks Inf. Secur. (2009). https://doi.org/10.5120/705-989

58. H.A. Abdul-Ghani, D. Konstantas, A comprehensive study of security and privacy guidelines, threats, and countermeasures: an IoT perspective. J. Sens. Actuator Netw. (2019). https://doi.org/10.3390/jsan8020022

59. T. Yaqoob, H. Abbas, M. Atiquzzaman, Security vulnerabilities, attacks, countermeasures, and regulations of networked medical devices-a review. IEEE Commun. Surv. Tutorials (2019). https://doi.org/10.1109/COMST.2019.2914094

60. G. Hernandez, O. Arias, D. Buentello, Y. Jin, Smart Nest Thermostat: A Smart Spy in Your Home, Black Hat USA (2014)

61. R.K. Sharma, D.B. Rawat, Advances on security threats and countermeasures for cognitive radio networks: a survey. IEEE Commun. Surv. Tutorials (2015). https://doi.org/10.1109/COMST.2014.2380998

62. L. Lilien, B. Bhargava, A scheme for privacy-preserving data dissemination, IEEE Trans. Syst. Man, Cybern.—Part A Syst. Humans. (2006). https://doi.org/10.1109/tsmca.2006.871655

63. J. Galbally, A new Foe in biometrics: a narrative review of side-channel attacks. Comput. Secur. (2020). https://doi.org/10.1016/j.cose.2020.101902

64. A. Saeed, S.A. Hussain, P. Garraghan, Cross-VM network channel attacks and countermeasures within cloud computing environments. IEEE Trans. Dependable Secur. Comput. (2020). https://doi.org/10.1109/TDSC.2020.3037022

65. Z. Wang, L. Sun, H. Zhu, Defining social engineering in cybersecurity. IEEE Access (2020). https://doi.org/10.1109/ACCESS.2020.2992807

66. F. Salahdine, N. Kaabouch, Social engineering attacks: a survey. Futur. Internet. (2019). https://doi.org/10.3390/FI11040089

67. X. Ai, H. Chen, K. Lin, Z. Wang, J. Yu, Nowhere to hide: efficiently identifying probabilistic cloning attacks in large-scale rfid systems. IEEE Trans. Inf. Forensics Secur. (2021). https://doi.org/10.1109/TIFS.2020.3023785

68. W. Huang, Y. Zhang, Y. Feng, ACD: An adaptable approach for RFID cloning attack detection. Sensors (Switzerland). (2020). https://doi.org/10.3390/s20082378

69. F. Laurenţiu Ţiplea, C. Andriesei, C. Hristea, Security and privacy of PUF-based RFID systems, in Cryptogr.—Recent Adv. Futur. Dev. [Working Title], (2020). https://doi.org/10.5772/intechopen.94018

70. M.A. Aladaileh, M. Anbar, I.H. Hasbullah, Y.W. Chong, Y.K. Sanjalawe, Detection techniques of distributed denial of service attacks on software-defined networking controller-a review. IEEE Access (2020). https://doi.org/10.1109/ACCESS.2020.3013998

71. L. Fang, B. Zhao, Y. Li, Z. Liu, C. Ge, W. Meng, Countermeasure based on smart contracts and AI against DoS/DDoS attack in 5G circumstances. IEEE Netw. (2020). https://doi.org/10.1109/MNET.021.1900614

72. S. Loureiro, Security misconfigurations and how to prevent them. Netw. Secur. (2021). https://doi.org/10.1016/S1353-4858(21)00053-2

73. A. Sołtysik-Piorunkiewicz, M. Krysiak, The cyber threats analysis for web applications security in Industry 4.0, in Studies in Computational Intelligence (2020). https://doi.org/10.1007/978-3-030-40417-8_8

74. Z. Cui, Y. Zhao, Y. Cao, X. Cai, W. Zhang, J. Chen, Malicious code detection under 5G HetNets based on a multi-objective RBM model. IEEE Netw. (2021). https://doi.org/10.1109/MNET.011.2000331

75. B. Vignau, R. Khoury, S. Hallé, A. Hamou-Lhadj, The evolution of IoT Malwares, from 2008 to 2019: Survey, taxonomy, process simulator and perspectives. J. Syst. Archit. (2021). https://doi.org/10.1016/j.sysarc.2021.102143

76. A.D. Raju, I. AbuAlhaol, R.S. Giagone, Y. Zhou, H. Shengqiang., A survey on cross-architectural IoT Malware threat hunting, IEEE Access (2021). https://doi.org/10.1109/access.2021.3091427.

77. H.A. Khan, N. Sehatbakhsh, L.N. Nguyen, M. Prvulovic, A. Zajić, Malware detection in embedded systems using neural network model for electromagnetic side-channel signals. J. Hardw. Syst. Secur. (2019). https://doi.org/10.1007/s41635-019-00074-w

78. M.Q. Li, B.C.M. Fung, P. Charland, S.H.H. Ding, I-MAD: Interpretable malware detector using Galaxy transformer. Comput. Secur. (2021). https://doi.org/10.1016/j.cose.2021.102371

79. B.A.S. Al-rimy, M.A. Maarof, S.Z.M. Shaid, Ransomware threat success factors, taxonomy, and countermeasures: a survey and research directions. Comput. Secur. (2018). https://doi.org/10.1016/j.cose.2018.01.001

80. A. Huseinović, S. Mrdović, K. Bicakci, S. Uludag, A survey of denial-of-service attacks and solutions in the smart grid. IEEE Access (2020). https://doi.org/10.1109/ACCESS.2020.302 6923

81. S. Ramesh, C. Yaashuwanth, B.A. Muthukrishnan, Machine learning approach for secure communication in wireless video sensor networks against denial-of-service attacks. Int. J. Commun. Syst. (2020). https://doi.org/10.1002/dac.4073

82. W. Zhao, S. Yang, X. Luo, On threat analysis of IoT-based systems: a survey, in Proc.—2020 IEEE International Conference Smart Internet Things, SmartIoT 2020 (2020). https://doi.org/10.1109/SmartIoT49966.2020.00038

83. U. Sabeel, S. Maqbool, Categorized security threats in the wireless sensor networks: countermeasures and security management schemes. Int. J. Comput. Appl. (2013). https://doi.org/10.5120/10718-5262

84. Y. Gao, B.G. Doan, Z. Zhang, S. Ma, J. Zhang, A. Fu, S. Nepal, H. Kim, Backdoor attacks and countermeasures on deep learning: a comprehensive review (2020). http://arxiv.org/abs/2007.10760

85. J.S. Cho, S.S. Yeo, S.K. Kim, Securing against brute-force attack: a hash-based RFID mutual authentication protocol using a secret value. Comput. Commun. (2011). https://doi.org/10.1016/j.comcom.2010.02.029

86. L. James, D. E.D., Technique to Thwart Brute-Force Attack : A Survey, Int. J. Sci. Res. Sci. Eng. Technol. (2020). https://doi.org/10.32628/ijsrset207139

87. S.H. Ahmadinejad, S. Jalili, M. Abadi, A hybrid model for correlating alerts of known and unknown attack scenarios and updating attack graphs. Comput. Netw. (2011). https://doi.org/10.1016/j.comnet.2011.03.005

88. L.J. Gonzalez-Soler, M. Gomez-Barrero, L. Chang, A. Perez-Suarez, C. Busch, Fingerprint presentation attack detection based on local features encoding for unknown attacks. IEEE Access. (2021). https://doi.org/10.1109/ACCESS.2020.3048756

89. S. Chakraborty, A comparison study of computer virus and detection techniques. Res. J. Eng. Technol. (2017). https://doi.org/10.5958/2321-581x.2017.00008.3

90. A. Belous, V. Saladukha, Viruses, hardware and software trojans: attacks and countermeasures (2020). https://doi.org/10.1007/978-3-030-47218-4

91. M.M. Naeem, I. Hussain, M.M. Saad Missen, A survey on registration hijacking attack consequences and protection for session initiation protocol (SIP), Comput. Netw. (2020). https://doi.org/10.1016/j.comnet.2020.107250

92. S. Sahoo, J.C.H. Peng, S. Mishra, T. Dragicevic, Distributed screening of hijacking attacks in DC microgrids. IEEE Trans. Power Electron. (2020). https://doi.org/10.1109/TPEL.2019.2957071

93. C. Basile, D. Canavese, L. Regano, P. Falcarin, B. De Sutter, A meta-model for software protections and reverse engineering attacks. J. Syst. Softw. (2019). https://doi.org/10.1016/j.jss.2018.12.025

94. A.M.H. Al-Hakimi, A.B.M. Sultan, A.A.A. Ghani, N.M. Ali, N.I. Admodisastro, Hybrid obfuscation technique to protect source code from prohibited software reverse engineering. IEEE Access (2020). https://doi.org/10.1109/ACCESS.2020.3028428

95. Q. Alasad, J.S. Yuan, P. Subramanyan, Strong logic obfuscation with low overhead against IC reverse engineering attacks. ACM Trans. Des. Autom. Electron. Syst. (2020). https://doi.org/10.1145/3398012

96. S. Charles, P. Mishra, A survey of network-on-chip security attacks and countermeasures. ACM Comput. Surv. (2021). https://doi.org/10.1145/3450964

97. C.Y. Yeh, E.W. Knightly, Eavesdropping in massive MIMO: new vulnerabilities and countermeasures. IEEE Trans. Wirel. Commun. (2021). https://doi.org/10.1109/TWC.2021.307 4941

98. I. Sutherland, G. Davies, A. Blyth, Malware and steganography in hard disk firmware. J. Comput. Virol. (2011). https://doi.org/10.1007/s11416-010-0149-x
99. A. Mirian, J. DeBlasio, S. Savage, G.M. Voelker, K. Thomas, Hack for Hire: exploring the emerging market for account hijacking, in Web Conference 2019—Proceedings World Wide Web Conference WWW 2019 (2019). https://doi.org/10.1145/3308558.3313489
100. S.S. Tirumala, H. Sathu, V. Naidu, Analysis and prevention of account hijacking based INCIDENTS in cloud environment, in Proceedings—2015 14th International Conference on Information Technology ICIT 2015 (2016). https://doi.org/10.1109/ICIT.2015.29
101. S.A. Islam, L.K. Sah, S. Katkoori, High-level synthesis of key-obfuscated RTL iP with design lockout and camouflaging. ACM Trans. Des. Autom. Electron. Syst. (2021). https://doi.org/10.1145/3410337
102. M. El Beqqal, M. Azizi, Review on security issues in RFID systems. Adv. Sci. Technol. Eng. Syst. (2017). https://doi.org/10.25046/aj020624
103. S. (Sy) Banerjee, T.A. Hemphill, P. Longstreet, Is IOT a threat to consumer consent? The Perils of wearable devices health data exposure. SSRN Electron. J. (2017). https://doi.org/10.2139/ssrn.3038872
104. R. Richardson, M. North, Ransomware: evolution, mitigation and prevention. Int. Manag. Rev. (2017)
105. X. Fu, Y. Gao, B. Luo, X. Du, M. Guizani, Security threats to Hadoop: data leakage attacks and investigation. IEEE Netw. (2017). https://doi.org/10.1109/MNET.2017.1500095NM
106. T. Aziz, E. Haq, Security challenges facing IoT layers and its protective measures. Int. J. Comput. Appl. (2018). https://doi.org/10.5120/ijca2018916607
107. P. Martins, L. Sousa, A. Mariano, A survey on fully homomorphic encryption: an engineering perspective. ACM Comput. Surv. (2017). https://doi.org/10.1145/3124441
108. Q. Xiao, C. Boulet, T. Gibbons, RFID security issues in military supply chains, in Proceedings—Second International Conference on Availability, Reliability and Security ARES 2007 (2007). https://doi.org/10.1109/ARES.2007.127
109. A. Mustafa, B. Poudel, A. Bidram, H. Modares, Detection and mitigation of data manipulation attacks in AC microgrids. IEEE Trans. Smart Grid. (2020). https://doi.org/10.1109/TSG.2019.2958014
110. P.M. Shakeel, S. Baskar, H. Fouad, G. Manogaran, V. Saravanan, C.E. Montenegro-Marin, Internet of things forensic data analysis using machine learning to identify roots of data scavenging. Futur. Gener. Comput. Syst. (2021). https://doi.org/10.1016/j.future.2020.10.001
111. A.K. Singh, B.D.K. Patro, Security attacks on RFID and their countermeasures (2021). https://doi.org/10.1007/978-981-16-0980-0_49
112. K. Schramm, T. Wollinger, C. Paar, A new class of collision attacks and its application to DES, Lect. Notes Comput. Sci. (Including Subser. Lect. Notes Artif. Intell. Lect. Notes Bioinformatics). (2003). https://doi.org/10.1007/978-3-540-39887-5_16
113. P.P. Pittalia, A comparative study of Hash algorithms in cryptography, Int. J. Comput. Sci. Mob. Comput. (2019)
114. Cloud computing: implementation, management, and security, Choice Rev. Online. (2010). https://doi.org/10.5860/choice.48-0915
115. A. Satpathy, S.K. Addya, A.K. Turuk, B. Majhi, G. Sahoo, Crow search based virtual machine placement strategy in cloud data centers with live migration. Comput. Electr. Eng. (2018). https://doi.org/10.1016/j.compeleceng.2017.12.032
116. B. Grobauer, T. Walloschek, E. Stöcker, Understanding cloud computing vulnerabilities. IEEE Secur. Priv. (2011). https://doi.org/10.1109/MSP.2010.115
117. J. Wu, Z. Lei, S. Chen, W. Shen, An access control model for preventing virtual machine escape attack. Futur. Internet. (2017). https://doi.org/10.3390/fi9020020
118. Y. Dong, Z. Lei, An access control model for preventing virtual machine hopping attack. Futur. Internet. (2019). https://doi.org/10.3390/fi11030082
119. N. Subramanian, A. Jeyaraj, Recent security challenges in cloud computing. Comput. Electr. Eng. (2018). https://doi.org/10.1016/j.compeleceng.2018.06.006

120. C. Baumann, M. Naslund, C. Gehrmann, O. Schwarz, H. Thorsen, A high assurance virtu-alization platform for ARMv8, in EUCNC 2016—European Conference on Networks and Communications, 2016. https://doi.org/10.1109/EuCNC.2016.7561034
121. L. Hong, H.C. Yong, Q.H. Zhang, The survey of RFID attacks and defenses, in 2012 International Conference on Wireless Communications Networking Mobile Computing WiCOM 2012 (2012). https://doi.org/10.1109/WiCOM.2012.6478720
122. M.M. Ahemd, M.A. Shah, A. Wahid, IoT security: a layered approach for attacks & defenses, in 2017 International Conference on Communication Technologies ComTech 2017 (2017). https://doi.org/10.1109/COMTECH.2017.8065757
123. A. Mitrokotsa, M.R. Rieback, A.S. Tanenbaum, Classifying RFID attacks and defenses. Inf. Syst. Front. (2010). https://doi.org/10.1007/s10796-009-9210-z
124. K. Bu, M. Weng, Y. Zheng, B. Xiao, X. Liu, You can clone but you cannot hide: A survey of clone prevention and detection for RFID. IEEE Commun. Surv. Tutorials. (2017). https://doi.org/10.1109/COMST.2017.2688411
125. A. Kumar, A.K. Jain, M. Dua, A comprehensive taxonomy of security and privacy issues in RFID. Complex Intell. Syst. (2021). https://doi.org/10.1007/s40747-021-00280-6
126. M. Roland, J. Langer, J. Scharinger, Practical attack scenarios on secure element-enabled mobile devices, in: Proceedings of the 4th International Workshop on Near Field Communication NFC 2012 (2012). https://doi.org/10.1109/NFC.2012.10
127. L. Francis, G. Hancke, K. Mayes, K. Markantonakis, Practical relay attack on contactless transactions by using NFC mobile phones. Cryptol. Inf. Secur. Ser. (2012). https://doi.org/10.3233/978-1-61499-143-4-21
128. G.P. Hancke, M.G. Kuhn, Attacks on time-of-flight distance bounding channels, in WiSec'08 Proceedings of the 1st ACM Conference on Wireless Network Security, 2008. https://doi.org/10.1145/1352533.1352566
129. S. Akter, S. Chellappan, T. Chakraborty, T.A. Khan, A. Rahman, A.B.M. Alim Al Islam, Man-in-the-middle attack on contactless payment over NFC communications: design, implementation, experiments and detection. IEEE Trans. Dependable Secur. Comput. (2020). https://doi.org/10.1109/tdsc.2020.3030213
130. E. Haselsteiner, K. Breitfuß, Security in near field communication (NFC) strengths and weaknesses. Semiconductors (2006)
131. C.H. Chen, I.C. Lin, C.C. Yang, NFC attacks analysis and survey, in Proceedings—2014 8th International Conferences Innovative Mobile and Internet Services in Ubiquitous Computing:. IMIS 2014 (2014). https://doi.org/10.1109/IMIS.2014.66
132. V. Ashktorab, S. Taghizadeh Reza, Security threats and countermeasures in Bluetooth-enabled systems. Int. J. Appl. or Innov. Eng. Manag. (2012)
133. N.B.N. Ibn Minar, Bluetooth security threats and solutions: a survey, Int. J. Distrib. Parallel Syst. (2012). https://doi.org/10.5121/ijdps.2012.3110
134. M. Chernyshev, C. Valli, M. Johnstone, Revisiting urban War Nibbling: mobile passive discovery of classic Bluetooth devices using Ubertooth One. IEEE Trans. Inf. Forensics Secur. (2017). https://doi.org/10.1109/TIFS.2017.2678463
135. H.J. Tay, J. Tan, P. Narasimhan, A survey of security vulnerabilities in Bluetooth low energy Beacons. Minor (2016)
136. A. Klein, Attacks on the RC4 stream cipher, Des. Codes, Cryptogr. (2008). https://doi.org/10.1007/s10623-008-9206-6
137. M. Caneill, J. Gilis, Attacks against the WiFi protocols WEP and WPA. Journal (2010)
138. M. Beck, E. Tews, Practical attacks against WEP and WPA, in Proceedings 2nd ACM Conference on Wireless Network Security WiSec'09, 2009. https://doi.org/10.1145/1514274.1514286.
139. C. Schmitt, T. Kothmayr, W. Hu, B. Stiller, Two-way authentication for the Internet-of-Things (2017). https://doi.org/10.1007/978-3-319-53472-5_2
140. M. Morii, Y. Todo, Cryptanalysis for RC4 and breaking WEP/WPA-TKIP. IEICE Trans. Inf. Syst. (2011). https://doi.org/10.1587/transinf.E94.D.2087

141. A.K. Kyaw, F. Sioquim, J. Joseph, Dictionary attack on Wordpress: security and forensic analysis, in 2015 2nd International Conference on Information Security and Cyber Forensics, InfoSec 2015 (2016). https://doi.org/10.1109/InfoSec.2015.7435522

142. L. Atzori, A. Iera, G. Morabito, The Internet of Things: a survey. Comput. Netw. (2010). https://doi.org/10.1016/j.comnet.2010.05.010

143. J. Gubbi, R. Buyya, S. Marusic, M. Palaniswami, Internet of Things (IoT): a vision, architectural elements, and future directions. Futur. Gener. Comput. Syst. (2013). https://doi.org/10.1016/j.future.2013.01.010

144. Q. Jing, A.V. Vasilakos, J. Wan, J. Lu, D. Qiu, Security of the Internet of Things: perspectives and challenges. Wirel. Netw. (2014). https://doi.org/10.1007/s11276-014-0761-7

145. C. Karlof, N. Sastry, D. Wagner, TinySec: a link layer security architecture for wireless sensor networks, in SenSys'04—Proceedings Second international conference on Embedded networked sensor System, 2004

146. C. Perera, A. Zaslavsky, P. Christen, D. Georgakopoulos, Context aware computing for the internet of things: a survey. IEEE Commun. Surv. Tutorials. (2014). https://doi.org/10.1109/SURV.2013.042313.00197

147. C. Zhen, H. Jianbin, C. Zhong, X. Maoxing, Z. Xia, Feedback: towards dynamic behavior and secure routing for wireless sensor networks, in Proceedings—International Conference on Advanced Information Networking and Applications AINA (2006). https://doi.org/10.1109/AINA.2006.179.

148. C. Modi, D. Patel, B. Borisaniya, H. Patel, A. Patel, M. Rajarajan, A survey of intrusion detection techniques in cloud. J. Netw. Comput. Appl. (2013). https://doi.org/10.1016/j.jnca.2012.05.003

149. C. Kolias, G. Kambourakis, A. Stavrou, J. Voas, DDoS in the IoT: Mirai and other botnets, Computer (Long. Beach. Calif) (2017). https://doi.org/10.1109/MC.2017.201.

150. Y. Liu, C. Cheng, T. Gu, T. Jiang, X. Li, A lightweight authenticated communication scheme for smart grid. IEEE Sens. J. (2016). https://doi.org/10.1109/JSEN.2015.2489258

151. Š. Bahtiyar, M. Ufuk Çağlayan, Extracting trust information from security system of a service, J. Netw. Comput. Appl. (2012). https://doi.org/10.1016/j.jnca.2011.10.002.

152. A. Akhunzada, A. Gani, N.B. Anuar, A. Abdelaziz, M.K. Khan, A. Hayat, S.U. Khan, Secure and dependable software defined networks. J. Netw. Comput. Appl. (2016). https://doi.org/10.1016/j.jnca.2015.11.012

153. C. Stergiou, K.E. Psannis, B.G. Kim, B. Gupta, Secure integration of IoT and cloud computing. Futur. Gener. Comput. Syst. (2018). https://doi.org/10.1016/j.future.2016.11.031

154. K. Lee, D. Murray, D. Hughes, W. Joosen, Extending sensor networks into the cloud using Amazon web services, in 2010 IEEE International Conference Networked Embedded Systems for Enterprise Applications NESEA 2010, 2010. https://doi.org/10.1109/NESEA.2010.5678063

155. S. Subashini, V. Kavitha, A survey on security issues in service delivery models of cloud computing. J. Netw. Comput. Appl. (2011). https://doi.org/10.1016/j.jnca.2010.07.006

156. T. Bhattasali, R. Chaki, N. Chaki, Secure and trusted cloud of things, in 2013 Annual IEEE India Conference INDICON 2013 (2013). https://doi.org/10.1109/INDCON.2013.6725878.

157. E. Shi, Y. Niu, M. Jakobsson, R. Chow, Implicit authentication through learning user behavior, in Lect. Notes Comput. Sci. (Including Subser. Lect. Notes Artif. Intell. Lect. Notes Bioinformatics) (2011). https://doi.org/10.1007/978-3-642-18178-8_9

158. S. Fremdt, R. Beck, S. Weber, Does cloud computing matter? An analysis of the cloud model software-as-a-service and its impact on operational agility, in Proceedings of the Annual Hawaii International Conference on System Sciences, 2013. https://doi.org/10.1109/HICSS.2013.182.

159. A. Ukil, S. Bandyopadhyay, A. Pal, IoT-privacy: to be private or not to be private. Proc.—IEEE INFOCOM (2014). https://doi.org/10.1109/INFCOMW.2014.6849186

160. H. Jiang, C. Cai, X. Ma, Y. Yang, J. Liu, Smart home based on WiFi sensing: A survey. IEEE Access (2018). https://doi.org/10.1109/ACCESS.2018.2812887

161. P. Faruki, A. Bharmal, V. Laxmi, V. Ganmoor, M.S. Gaur, M. Conti, M. Rajarajan, Android security: a survey of issues, malware penetration, and defenses. IEEE Commun. Surv. Tutorials (2015). https://doi.org/10.1109/COMST.2014.2386139
162. Q. Li, B. Sun, M. Chen, H. Dong, Detection malicious Android application based on simple-Dalvik intermediate language. Neural Comput. Appl. (2019). https://doi.org/10.1007/s00521-018-3726-4
163. J. Huang, X. Zhang, L. Tan, P. Wang, B. Liang, AsDroid: detecting stealthy behaviors in Android applications by user interface and program behavior contradiction. Proc.—Int. Conf. Softw. Eng. (2014). https://doi.org/10.1145/2568225.2568301
164. J. Qiu, J. Zhang, W. Luo, L. Pan, S. Nepal, Y. Xiang, A survey of Android Malware detection with deep neural models. ACM Comput. Surv. (2021). https://doi.org/10.1145/3417978
165. V. Sihag, M. Vardhan, P. Singh, A survey of android application and malware hardening. Comput. Sci. Rev. (2021). https://doi.org/10.1016/j.cosrev.2021.100365
166. S.R. Steinhubl, E.D. Muse, E.J. Topol, The emerging field of mobile health. Sci. Transl. Med. (2015). https://doi.org/10.1126/scitranslmed.aaa3487

Chapter 4
Digital Forensics in Internet of Things

As IoT technology becomes an integral part of everyday life, enhancing productivity for businesses through automation, it should come as no surprise that attackers would seek to exploit these systems and the services they provide for profit. Through the Internet, attacks are capable of launching a number of diverse malicious actions such as Distributed Denial of Service (DDoS), Scanning/Probing, Keylogging, Malware proliferation, E-mail spamming, Click fraud, Phishing, Identity theft, and more [1, 2]. As such, the need for reliable methods of investigation, which can be used for identification of security incidents, reconstruction of events and attribution is evident. As a result, the discipline of Digital Forensics was developed and continues evolving to cope with the composite nature of IoT environments resulting in what we call IoT forensics. This section is an introduction to digital forensics and its concepts.

The main objectives of this chapter are as follows:

1. To understand the digital forensic process.
2. To discuss the history of digital forensics and its related disciplines.
3. To learn the digital forensic investigation steps.
4. To discuss the various types of cyber-crime and digital evidence.

4.1 What Is Digital Forensic?

In the center of all forensic disciplines and subdisciplines lies Locard's Exchange Principle [3, 4] which states that any contact that the criminal has had with the crime scene, will leave backtraces of that interaction. Thus, the purpose of forensics has always been to identify these traces, in order to assist a criminal investigation in apprehending a criminal.

Electronic Supplementary Material The online version of this chapter (https://doi.org/10.1007/978-3-030-89025-4_4) contains supplementary material, which is available to authorized users.

M. Abdel-Basset et al., *Deep Learning Techniques for IoT Security and Privacy*,
Studies in Computational Intelligence 997,
https://doi.org/10.1007/978-3-030-89025-4_4

Table 4.1 McKemmish digital forensics stages

Order	Step	Description
1	Identification	Characterizes the prerequisites for evidence administration, realizing it is current while informed about the corresponding kind, position, and format
2	Preservation	Emphasis guaranteeing that the evidential data stays untouched or modified slightly
3	Analysis	Interprets and transforms the data collected into evidence
4	Presentation	Presents evidence to the judges in terms of offering specialist proof on the analysis of the evidence

As criminals have expanded their activities into the Internet and make use of computers to commit their crimes, so has law enforcement and forensics evolved in order to keep up with the criminals, giving rise to a new discipline called digital forensics. There have been multiple definitions for digital forensics, each with its own merits, proposed by different organizations and viewing the fields from different perspectives [1]. One such popular definition by McKemmish (1999) [5] is that: *digital forensics is the process of identifying, preserving, analyzing, and presenting digital evidence in a manner that is legally acceptable.* Thus, based on McKemmish, the process starts with an investigator identifying possible sources of evidence and the type of data. Next, the investigator needs to preserve the crime scene by ensuring that data isn't altered during collection or is changed as little as possible. During analysis, the gathered data is processed and analyzed to discover true evidence and make inferences about the case, which are then presented often in a court of law. Where the role of each step is clearly described as presented in Table 4.1. In other words, the fundamental objective of digital forensics is to find evidence to answer five W-started and How (5WH) questions, which involve What happened, who was engaged, When did it happen, Where did it happen, Why did that happen, and How an incident happened. Answering these questions leads to verifying or negating accusations of an incident [6–8].

Another definition for digital forensics, given by National Institute of Standards and Technology (NIST) [9] as *"Digital forensics is the field of forensic science that is concerned with retrieving, storing and analyzing electronic data that can be useful in criminal investigations. This includes information from computers, hard drives, mobile phones, and other data storage devices.".* This definition identifies four slightly different steps of digital forensics: Collection, Examination, Analysis, and Reporting [10]. Where the role of each step is clearly described as presented in Table 4.2.

There is some overlap between the stages by McKemmish and NIST, as given in Fig. 4.1. In Fig. 4.2, the process of digital forensics is defined in terms of eight successive steps including Identification, preservation, collection, examination, analysis, interpretation, documentation, and presentations. As computers rely on several specialized subsystems to function (i.e., Hard disk (HD), Random Access Memory (RAM), Network Interface Card (NIC), etc.), attackers may target any subsystem,

Table 4.2 Steps of digital forensics according to NIST

Order	Steps	Description
1	Collection	This step aims to detect any possible sources of data related to the confrontation and then to tag and record them. Next, the data situated in these sources ought to be obtained whilst maintaining the sources' integrity
2	Examination	This step entails evaluating the obtained data from the previous step (i.e., Collection) and extricating the data related to the incident whilst conserving its integrity
3	Analysis	This step entails exploring the information mined by the investigation to answer the 5WH questions and/or decide that no or incomplete decision can be taken
4	Reporting	This step describes the process of formulating and presenting the practice, techniques and devices employed in the investigation together with the findings and outcomes gained from the analysis step

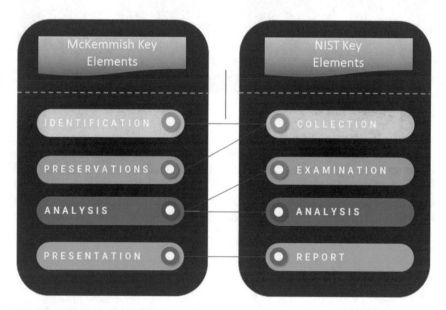

Fig. 4.1 McKemmish NIST key elements

depending on the purpose of the attack. Thus, in digital forensics, there are multiple methods for:

- Finding data upon computer systems.
- Retrieving removed, encrypted, or harmed file information.
- Scrutinizing real-time action.
- Identifying contraventions of business policy.

Fig. 4.2 Digital forensic life cycle

Evidence acquired during a digital forensic investigation may be necessary for a broad scope of digital crimes and misuse. The collected information may be used by law enforcement, assisting in arrests and prosecution, but it can also be used by companies for termination of employment due to sabotage or misuse of corporate systems, or even for counter-terrorism in order to prevent future illegal activities.

4.2 Digital Evidence

The analysis phase of an investigation identifies evidence in the data. **Digital Evidence**, interchangeably identified as electronic evidence, is defined as data or information often available in digital layout, which can be exploited to provide proof or disclose the truth on the subject of a cyber-crime and can be count on and used in a law court. Digital evidence can take many forms but, can be thought of as any data remaining prone to human interference or not, which could be extricated from a system such as computers, mobile phones, and embedded devices. It must be in human-comprehensible layout or having the ability to be understood by somebody having experience in the topic, in order to be used in legal procedures.

4.2.1   Digital Forensics and Other Related Disciplines

As digital evidence can be located anywhere inside a computer, digital forensic activities vary in their complexity and area of application. Examples of digital forensic activities include recovering evidence from a formatted hard drive, performing investigation after multiple users have taken over a system, or recovering deleted emails/files. It should be noted that digital forensics differs from data recovery. Data recovery is employed to retrieve files that were removed accidentally or lost/corrupted through a power outage, while digital forensics focuses primarily on finding evidence.

Forensics investigators habitually perform their missions as members of a work team, identified as the investigation triad [11]. First the threat evaluation as well as risk management, checks and validates the integrity of longstanding computer terminals and network servers. The tasks of incident response and detection of network intrusion use automated tools to detect intruders and monitors the logs of the network firewall. Finally, digital investigations handle investigations and perform forensics analysis of systems believed to contain evidence. Figure 4.3 gives the investigation triad as three sides of a triangle, all working together.

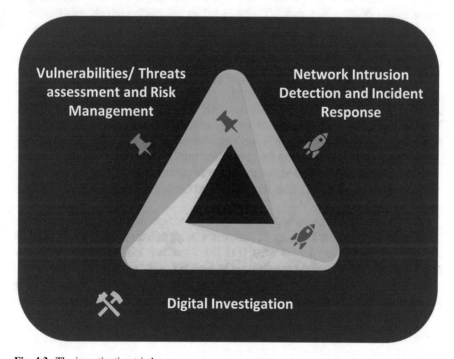

Fig. 4.3 The investigation triad

4.2.2   A Brief History of Digital Forensics

At first, digital forensics was a misunderstood domain, with minimal space and resources being allocated for investigators to analyze digital data [12]. But as cyber threats started making their appearance, governments and law enforcement agencies started to take notice of this new discipline. One of the first conferences in the field was hosted by the FBI academy in 1993, "First International Conference on Computer Evidence" gathered the attention of 26 countries and was one of the first opportunities for practitioners from around the world to exchange ideas.

One of the first education programs focusing on digital forensics appeared in the early 1990s when the International Association of Computer Investigative Specialists (IACIS) launched training on software tools for digital forensic investigators. Gradually, various agencies started developing their own digital forensic software. The IRS designed search-warrant programs, used to prepare procedures for executing warrants and search/seizure [12, 13].

The first commercially available digital forensic software tool was developed by a group called ASR Data (www.asrdata.com). Designed for Macintosh machines, the tool was called Expert Witness and allowed searches that spanned an entire hard drive [14]. Software developed and maintained by the IRS, specifically the Criminal Investigation Division is iLook, a Linux-based software allowing an investigator to acquire a complete image of a computer system and quickly review its hard drive [15]. Finally, a popular commercial product is the AccessData Forensic Toolkit (FTK) which can be used to fully examine a computer, providing e-mail retrieval services and decrypting information found in the registry [13, 16].

As stated by McKemmish [17], results produced by a digital forensic process are often used to support or prove some event or activity, often of a legal nature. Thus, it is critical that the results of such an investigation be reliable and accurate, something that can be achieved through transparency. By ensuring that the entire process and results are reproducible under the same conditions, verification of the forensic process is achieved and confidence in the results is built. The resulting problem thus is as follows: how can we ensure the reliability and accuracy of the digital forensic process?

With technology changing rapidly and cybercrimes keeping pace, it becomes clear that current laws cannot keep up. In such cases, where statues either do not exist or cannot be applied, case law is used [11, 18]. Case law lets legitimate advocates apply preceding comparable case rulings to existing cases as an attempt to address ambiguity in laws. This, however, means that inspectors have to be comfortable with the latest court judgments on exploration and seizure in computerized environments.

4.2.3   Common Sources of Digital Evidence

In previous decades, before the emergence of the Internet of Things (IoT) and ubiquitous computing, the sources of digital evidence were few and easily discoverable. Such sources included PCs, laptops, routers, and servers. However, the assimilation of ubiquitous computing in everyday life has caused the number of sources of evidence to increase drastically [19]. Due to the nature of IoT devices being small in size, portable, and easily concealable, forensic investigators often have a hard time identifying possible sources of evidence [19]. Thus, in the IoT era, possible evidence sources may include cloud storage servers, client devices like Android or iOS, IoT devices like drones, smart TVs and smart meters, other cyber-physical elements. Having discussed the possible sources of evidence, it is prudent to talk about the different types of evidence and how they may be interpreted and utilized. Through digital forensics, practically any file can be recovered, examined, and used in an investigation. Some evidence includes e-mail which is arguably the most important type of digital evidence, images, videos, browser search history, cell phone calls history, hard drive content, and internet communication [20].

4.2.4   Understanding Law Enforcement Agency Investigations

Digital forensic investigations can be carries out by either security experts, working in the private sector, investigating security breaches for companies, or by law enforcement agents, investigating cyber-related crimes. Although, unlike investigations carried out in the private sector, law enforcement agents are bound to adhere to the laws which dictate proper procedure for investigations [11].

For example, evidence acquired in breach of the Telecommunications Interception Act 1979 [21, 22], which prohibits the access of telecommunications without the knowledge of the participants, are rendered inadmissible in court. For search and seizure operations, investigators need to work within the guidelines of their warrant, and either extend it or acquire a new one if necessary. Furthermore, investigators need to be able to build a criminal case, by combining all legally obtained evidence in such a way, that a cyder crime is linked to the defendant with as little doubt as possible [11].

In the United States, the Computer Fraud and Abuse Act 1986 [23] is the primary federal law covering cybercrime incidents. Since it was introduced, it has been amended several times, with an example being the use of the Patriot Act to amend the definition of "protected computers" to include any and all systems either situated within the US or outside of it, that may affect commerce or communications of the US [24].

Regardless of the scene, any forensic investigation needs to include the following objectives:

- Collect useful evidence.
- Do not interrupt business processes.
- Ensure that evidence has a positive impact on outcomes and legal action.
- Assist any potential investigation of crimes and persuade adversaries to avoid further actions against the organization.
- Offer a procedure that has an acceptable cost.

4.3   Major Areas of Investigation for Digital Forensics

Primarily, digital forensics includes the detection and retrieval of information utilizing a variety of techniques and kits accessible by the detective or investigator. However, digital forensics may be applied to a number of diverse scenes, each requiring different methods. In this view, the forensics investigations involve, but are not restricted to the following:

- Data recovery [12]: Investigation and recovery of information that might have been removed, altered to a variety of file formats, or made concealed. It involves automated tools that access filesystems like FAT and NTFS and recovering files that may have been deleted, but still exist in the unallocated space of the Hard Drive.
- Identity theft [25]: Numerous dishonest behaviors fluctuating from nicked credit cards to phony social media accounts typically include some kind of individuality burglary. The proliferation of e-banking, e-commerce, and the digitization of various government services (social security numbers), have made identity theft much more efficient. Attacks that result in identity theft include phishing and pharming.
- Malware and ransomware investigations [26, 27]: Up till now, ransomware proliferated by worms and Trojans throughout local and public networks are considered as one of the greatest threats to corporations, governmental sectors, military, private businesses, and people. One infamous ransomware strain called CryptoLocker functioned by encrypting files in the target machine and then requesting payment to decrypt them. Ransomware Attacks have been costing victims a total of $20 billion by 2020. Malware can also be propagated to and by mobile IoT devices and stationary counterparts. One such scenario, with malware using the IoT to spread malware, was described by Ronen et al. [27]. The researchers showed that by hacking a Philips Hue light system and using the ZigBee protocol, a worm of their creation could infect neighboring IoT lamps, potentially infecting an entire city.
- Network and Internet investigations [28]: Probing DoS attacks and DDoS attacks and tracing the edited devices involving copiers, laptops, and files. Depending on the technique used, these attacks can be categorized into the volumetric, protocol, and application-based attacks. Detection methods include knowledge-based which rely on pre-defined rules, statistical methods which normally build a model for normal traffic (anomaly detection, NIDS), and machine learning

with classifiers such as Recurrent neural networks, convolution networks, and transformer networks.

- Email investigations: it involves source investigation, and IP tracking, content probing, and location investigation.
- Corporate espionage: A Variety Of enterprises are shifting from traditional hard copies of data and to cloud storage and services. Intrinsically, a digital trajectory is all the time enduring.
- Child pornography investigations: Unfortunately, the realism is that teenagers are extensively abused on the internet and contained by the world wide web. With the use of technology and highly skilled forensic analysts, investigations could be performed in taking down abuse bracelets by scrutinizing payment transactions, emails, images, browsing history, and even internet traffic.

4.4   Following Legal Processes

For a criminal investigation to start, there needs to be a starting point. This starting point can be either evidence, that someone discovered for some crime or a witness. Regardless, a witness or a victim reports a crime to the police, making an allegation. Following an allegation, police officers take statements and write a report about the crime [29, 30]. The report is then processed by management, whereupon it is decided whether to launch an investigation or log the information to a historical database where information about previous crimes is stored, called a blotter.

When an incident has been reported and an investigation is launched, one of the first people to arrive at the crime scene is the Digital Evidence First Responder (DEFR). The DEFR arrives at the scene of an incident and initially assesses the situation. Before any other action, precautions need to be taken, so that no information is lost due to sudden power-offs or malicious software. A very important action that needs to be performed at this stage, is establishing a chain of custody, which maintains who, where, how, why, and when interacted with the data before the investigators arrived and during the investigation [31]. Then, the DEFR acquires and preserves evidence [32].

The next stage of the investigation process relies on the actions of the Digital Evidence Specialist (DES). The DES has the technical skill to perform data analysis on the collected data from the previous stage [33]. Furthermore, one of the responsibilities of the DES, is to judge whether another specialist is to be called and assist with the investigation. Finally, to support facts about any identified evidence in a court of law, experts and witnesses are required to provide a written statement called an affidavit. An affidavit is legally binding and must include exhibits that support the claims asserted by it [11, 34, 35].

4.5 Types of Digital Evidence

Depending on their status, data may be categorized in one of three groups [32, 36]: active data, ambient data, and archive data. Active data include files that are, at the time of collection, normally stored and maintained in the storage. In other words, these files are easily accessible, without a need to re-construct them, although they may be otherwise protected (encrypted). Ambient data are files that have been deleted, and thus exist in the disk storage possibly fragmented. If the files were deleted from the computer device, accessing them may require scanning the disk storage in order for them to be reconstructed. Finally, archive data describe data that have been stored in backup data stores.

4.6 The Cyber Kill Chain

The concept "kill chain" is initially presented by the military for defining the phases an adversary follows to attack a goal. By 2011, Lockheed Martin published a study that defines the "Cyber Kill Chain" [37]. As with the military definition, Martin's study describes the phases applied by the attackers in the present cyber-attack's scenarios. Hypothetically, with the realization of those phases, defenders can better identify and stop attacks at each stage. By intercepting attackers at multiple stages of the kill chain, a defender increases their chances of shutting down the attackers' actions and denying them their objective. Since its introduction in 2011, a variety of editions of the "Cyber Kill Chain" have been proposed. In this section, we will discuss in what way the humanoid factor handles the earliest Martin's cyber kill chain (see Fig. 4.4).

First, remember that security consciousness is not anything more than a control similar to or anti-virus, passwords, encryption, or firewalls. The reason that renders security consciousness distinctive is that it pertains to and handles various level risks. Since security consciousness delivers the human aspects, individuals habitually believe it is not fitting to the Cyber Kill Chain. Nevertheless, this in fact is not true at all. In follow, the seven phases to the Cyber Kill Chain are discussed along with the means reliable personnel could assist counteract a cyber-attack per each phase.

- Reconnaissance: The attacker monitors and collects information about the victim before the concrete incident is launched. Several security specialists believe that there is nothing that the defenders could perform throughout this phase, however, that is not true. fairly often, cyber attackers search the internet to gather information regarding their expected victims. Besides, they might attempt to collect information through methods such as email contacts, phoning employees, etc. This is where trained and discipline could result in great effects. An experienced workforce is likely to realize that they might be a victim and reduce what they openly disclose. They have to verify individuals on the mobile prior to sharing any confidential information. They securely decide on and scrap confidential records

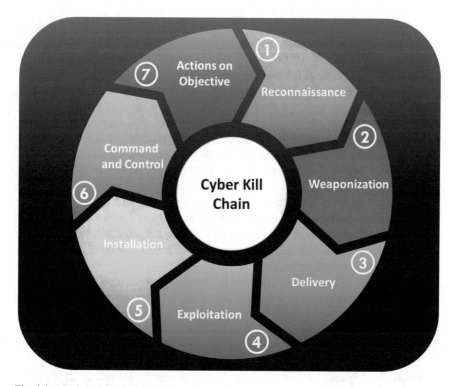

Fig. 4.4 Martin's cyber kill chain

and files. Although these actions may not entirely neutralize this stage, they can pose a great impression on the attacker's abilities.

- Weaponization: Having collected any necessary information from their reconnaissance, the attackers construct their attack in this stage. For instance, the attacker/intruder might design a contaminated file combined with a personalized phishing email, or else maybe the attacker designed novel fatigue of self-duplicating malware to be delivered over a USB drive. There are not many security mechanisms to improve security consciousness that influence or counteract per this phase, if not the attacker performs some restricted assessment on the expected victim.

- Delivery: In this stage, the prepared attack is transmitted to the expected target(s). For instance, distributing the contaminated USB drives at someplace, or transmitting the genuine phishing email. Whilst there is absolute technological engineering devoted to preventing this phase, individuals have a crucial responsibility. people are the first stage to detect and stop many of today's attacks, including modern or clientele attacks such as Spear Phishing, CEO Fraud, etc. Additionally, individuals could recognize and block attacks that nearly all technical tools unable to detect, like social engineering attacks. skilled personnel considerably diminishes that attack surface.

- Exploitation: This stage implies that the attack has been launched and the victim machine has been breached, with the selected exploit running on the system. Trained people ensure that the systems they are using are up to date. They make sure they have anti-virus functioning and allowed, and that the anti-virus has been updated with the latest attack fingerprints. An additional security measure is to make sure that any sensitive data are stored in secured systems, thus protecting them from exploits.
- Installation: The attackers set up malware at the machine of the attack target. Some 1 attacks do not involve malware, such as gathering login credentials or CEO Fraud attacks. Nevertheless, just the same as exploitation once the malware is being implicated, skilled and reliable personnel could assist guaranteeing they are utilizing protected devices that are efficient and come up with anti-virus empowered to stop lots of tries to install malwares. A major phase in identifying a contaminated system is exploring and discovering irregular behavior. Thus, the people that interact with computer systems that may be infected, are ideal for detecting any abnormalities.
- Command & Control: Once the system has been compromised, it is then managed remotely through a Command and Control (C&C) infrastructure. Usually, the infected machine will initiate the communication with the C&C to signal that the infection was successful, and then the attacker can issue commands. This is the reason why 'hunting', a process that involves looking for abnormal outbound activities, has become so popular.
- Actions On Objectives: After the cyber attackers gain access to the victim device or system, they later execute some activities to accomplish the targeted objectives. Motives significantly differ reliant on the attack player but may involve diplomatic, economic, or military benefits, which makes it complicated to specifically pinpoint these kinds of activities. Nevertheless, another time, this is where skilled personnel of Human Sensors implanted through a target can greatly enhance its capability to identify and act in response to a confrontation. Besides, secure actions will make it far more challenging for an effective opponent to revolve through the victim's system and accomplish their objectives. Behavior such as applying robust, distinctive passwords, validating individuals prior to communicating sensitive information, or securely resolving confidential information is just some of the countless manners that make the attacker's existence more problematic and cause them to be more liable to be distinguished.

The cyber kill chain model seeks to achieve two objectives. First, it could be applied for criminal intelligence with the aim to empower defensive abilities to be affiliated to the steps an enemy follows. Second, the analysis of intrusion, nevertheless, undertakes that the recognition of a cyber-attack is founded on an Indicator of compromise (IoC). As Soon as an IoC is discovered, then the analysis ought to begin from the step to which this IoC belongs and moving backward to previous steps as presumed to be fulfilled already. Hence, it is apparent that the cyber kill chain model and IoCs are likely to be utilized together in the investigation given that there is previous knowledge regarding a cyber-attack, which means that one or more

IoC must be discovered. Furthermore, IoCs cannot be considered a scheme to be employed for performing investigation hence they cannot guide it but just support it.

4.7 IoT Forensics

The IoT Forensics can be identified as a new subfield of the Digital Forensics. Nevertheless, whilst the standard digital forensics discipline has been existing in both academia and industry for a long time, IoT Forensics is a comparatively fresh field and still in its early phases. The objective of IoT Forensics is the same as with digital Forensics, which is to discover and obtain digital information in a legitimate and forensically sensible fashion. In other words, IoT forensics seeks to answer the 5WH questions in the context of IoT systems and constituent layers. In addition to gathering forensic data from physical IoT devices, the forensic data can also aggregate from local networks (e.g., a firewall or a router), edge or fog nodes, cloud datacenters [2]. Inspired by this fact, IoT Forensics can be taxonomized into five main subcategories as shown in Fig. 4.5. This illustrated taxonomy divides IoT forensics into IoT device forensics, Fog and Edge forensics, network forensics, application forensics, and cloud forensics.

The main discrepancy between IoT Forensics and digital forensics can be observed in terms of evidence sources. Different from standard digital forensics, in which the laptops, computers, smartphones, smartwatches, cloud servers are employed as a normal substance of inspection, the sources of evidence within IoT Forensics can

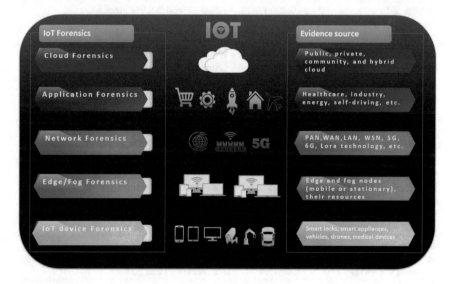

Fig. 4.5 Categories and/or levels of internet of things forensics

be much extra wide-spreading containing medical implantations in creatures, infant nursing systems, traffic lights, and In-Vehicle Infotainment (IVI) systems [38].

Preserving the security of overall IoT devices, communications, and storage within different layers of the IoT network is essentially difficult if not impossible to realize. If an incident happens, one of the first tasks that forensics professionals execute is to define the scope of the compromise. Nevertheless, different from standard IoT security procedures, the target of IoT forensics is not reducing the amount of incurred destruction, yet to distinguish the origin of attacks or the responsibilities of the various factions. The systematic comparison between different aspects of the two fields is reviewed in Table 4.3.

Despite the recent emergence of the concept of IoT forensics, yet it can be considered an extremely important topic for either the industrial or academic domains for many reasons. Firstly, the m*assive attack surfaces are* introduced by the dynamic nature of IoT environments. In this context, the recent advances in IoT have been greatly empowering the Machine-to-machine (M2M) technology at a wide scale of applications. However, the M2M technology is demonstrated as a typical example of chiefly vulnerable to connectivity cyber-attacks as it is usually connected with a huge number of nodes, embedded devices, servers. A similar IoT example can be

Table 4.3 Comparison between IoT security and IoT forensics

IoT security	IoT forensics
Delivers security insurance for physical and cybersecurity concerns	Determine and recreates the chain of incidents by examining physical and digital evidence cyber-physical context
Applies different security procedures to reduce the scale of the attack and avoid potential destruction	Applies investigative procedures for recognizing, capturing, preserving, and analyzing digital information
Real-time reply: applies a variety of techniques to tackle the threats throughout a live event	Post-mortem investigation: recognizes shortages following the occurrence of an incident or whilst the IoT system is inactive
Generalized: exploring for any potential damaging actions	Case-focused: rebuilding a provided criminal situation
Uninterrupted practice: stays on the alert for 24 h per day	Time-confined practice: following a crime is claimed to have taken place
Security training and practices use a set of security procedures, practices, and specifications, with the aim to realize a reliable IoT system and avoid imminent cyber-physical threats from going on	Forensic Readiness: fulfill the forensics needs and employs forensics standards, to be willing to carry out an investigation; takes weights to augment the forensic significance of the possible evidence, and reduce the number of consumed resources on the investigation
Indicates the legal state and legal facets in service legal accords concerning the security requirements and goals	Stipulate the legal state and legal facets in service legal accords concerning the forensics requirements and goals
The well-founded discipline of computer science	The fresh and unexplored discipline of Digital forensics

inferred for each IoT attack surface discussed in previous sections. Appropriately, IoT devices having a public crossing point are subjected to a bigger degree of risks for the reason that they can carry malware to the private network from public space with fewer protection [7]. Regularly realized incidents incorporate node tampering, data leakage, identity theft, taking control of Internet-connected devices, acquisition of cloud-based CCTV units, SQL injections, phishing, assurance interrelated fraud, cyberbullying, ransomware, and malware targeting specific appliances smart TVs, smartwatches.

Secondly, *emerging cyber-physical security and privacy threats* can be considered as a second reason that obligates the development of IoT forensics. with the dynamic and evolving nature of IoT technology, a new combination of more complex and dangerous attacks is likely to emerge with a new capability to overcome the current security solutions, thereby cyber-crimes now stride beyond the boundary of cyberspace and can jeopardize human lives or even communities. For instance, in January 2017, the United States Food and Drug Administration (FDA) announced a caution that specific pacemaker models are susceptible to hacking [11, 39]. Another vulnerability was detected in the portal login process of an LG smart vacuum cleaner, allowing a group of researchers to access a live video stream from inside the owner´s home [19]. Another example, when the CCTV units for monitoring (generally known as "nanny cams") were accessed and the footage freely offered to the community. In another way, Smart locks might be instructed to open when a certain device is identified by the building wireless network, letting a lawbreaker gain access to organizations, banks, somebody's home, or department. A scenario with great effects can be possible when a smart lock is trained to lock once a gas leak or fire is discovered [7]. This situation could be caused by a deliberate attack, or the corollary of a malfunction because of an inefficient construct and an absence of an adaptive system.

Thirdly, *Digital traces,* which can be defined as a fragment of information (hosted by IoT devices) that is capable to confirm or negate some assumption and can consequently, assist the forensics specialists in discovering answers for 5WH questions and empower rebuilding the original crime scene. For instance, digital traces might provide information concerning the time the alarms of the smart building were deactivated, and the time of opening a specific smart door. Likewise, the information collected from the carbon monoxide sensor might reveal the precise time and position at which the fire in the building was launched [40]. Wearables similar to smartwatches or health trackers can be employed to distinguish an individual through their biometric information [41]. Typical examples for such forensic items might involve temporarily stored image thumbnails and pieces of the camera flows, as well as cached events triggered by the sensors, and complete event logs stored in the application database. Undoubtedly, these records involve confidential private information about human's identity, location, actions, as well as general connections and narrative, and hence, have to be extracted and analyzed with distinct consideration to ethics and privacy of personal information.

4.8   Summary and Learnt Lessons

This chapter provides an in-depth discussion about the IoT security requirements, IoT threats, IoT vulnerabilities, Risks, and attacks providing a taxonomy for categorizing different threats, indicating the main IoT attack surfaces. In the nutshell, this chapter delivers a useful insight about different aspects of IoT security that could be pointed out as follow:

- In the later section of this chapter, we introduced the concept of digital forensics, its origins, its related domains, and its role in the IoT environment. Besides, the digital forensics investigation process is discussed along with the legal requirements that govern how an investigation is to be conducted in the context of IoT environment.
- Furthermore, the process of acquiring evidence from a cybercrime scene is extensively analyzed, then the cyber kill chain is discussed in detail. Finally, a detailed comparison is introduced to differentiate and distinguish IoT forensics from IoT security.
- The overabundance of challenges in the IoT Forensics signifies the absence of security in cyberspace. As A Result, researchers and forensics specialists should strive to discover and develop an innovative solution for precise gathering and maintenance of digital evidence. Legitimate entities, server providers, as well as device manufactures should collaboratively participate in removing the IoT security challenges.

References

1. N. Koroniotis, N. Moustafa, E. Sitnikova, Forensics and deep learning mechanisms for Botnets in Internet of Things: a survey of challenges and solutions. IEEE Access (2019). https://doi.org/10.1109/ACCESS.2019.2916717
2. B. Narwal, A.K. Mohapatra, K.A. Usmani, Towards a taxonomy of cyber threats against target applications. J. Stat. Manag. Syst. (2019). https://doi.org/10.1080/09720510.2019.1580907
3. The Forensic Laboratory Handbook Procedures and Practice (2011)
4. C.W. Muehlberger, P.L. Kirk, Crime investigation—physical evidence and the police laboratory. J. Crim. Law. Criminol. Police Sci. (1954). https://doi.org/10.2307/1139939
5. R. Mckemmish, What is forensic computing? Change (1999)
6. K. Tange, M. De Donno, X. Fafoutis, N. Dragoni, A systematic survey of Industrial Internet of Things security: requirements and fog computing opportunities. IEEE Commun. Surv. Tutorials (2020). https://doi.org/10.1109/COMST.2020.3011208
7. N. Akatyev, J.I. James, Evidence identification in IoT networks based on threat assessment. Futur. Gener. Comput. Syst. (2019). https://doi.org/10.1016/j.future.2017.10.012
8. C. Esposito, A. Castiglione, B. Martini, K.K.R. Choo, Cloud manufacturing: security, privacy, and forensic concerns. IEEE Cloud Comput. (2016). https://doi.org/10.1109/MCC.2016.79
9. K. Kent, S. Chevalier, T. Grance, H. Dang, Guide to Integrating Forensic Techniques into Incident Response (The National Institute of Standards and Technology, 2006).
10. A. Dimitriadis, N. Ivezic, B. Kulvatunyou, I. Mavridis, D4I—DIGITAL forensics framework for reviewing and investigating cyber attacks. Array (2020). https://doi.org/10.1016/j.array.2019.100015

11. M. Stoyanova, Y. Nikoloudakis, S. Panagiotakis, E. Pallis, E.K. Markakis, A survey on the Internet of Things (IoT) forensics: challenges, approaches, and open issues. IEEE Commun. Surv. Tutorials (2020). https://doi.org/10.1109/COMST.2019.2962586
12. Open University, A Brief History of Digital Forensics (Open University, 2019)
13. M.E. Busing, J.D. Null, K.A. Forcht, Computer forensics: the modern crime fighting tool. J. Comput. Inf. Syst. (2005). https://doi.org/10.1080/08874417.2006.11645889
14. P. Hawkins, Macintosh forensic analysis using OS X. Inf. Secur. (2001)
15. D. Hartanto, Analysis of mental models at criminal investigation division of the Indonesian National Police. Tech. Soc. Sci. J. (2020). https://doi.org/10.47577/tssj.v12i1.1575
16. Access Data, Forensic Toolkit (FTK) | AccessData. Forensic Toolkit (FTK) (2019)
17. R. McKemmish, When is digital evidence forensically sound? IFIP Int. Fed. Inf. Process. (2008). https://doi.org/10.1007/978-0-387-84927-0_1
18. V. Farhat, B. Mccarthy, R. Raysman, Cyber attacks : prevention and proactive responses. Pract. Law (2011)
19. S. Alabdulsalam, K. Schaefer, T. Kechadi, N.A. Le-Khac, Internet of things forensics—challenges and a case study. (2018). https://doi.org/10.1007/978-3-319-99277-8_3
20. D. Quick, K.K.R. Choo, Big forensic data reduction: digital forensic images and electronic evidence. Cluster Comput. (2016). https://doi.org/10.1007/s10586-016-0553-1
21. T. Round, M. Hyde, D. Power, Telecommunication devices for the deaf and the telecommunications (Interception) Act 1979. Aust. J. Law Soc. **13**(1997), 181–190 (1997)
22. N.M. Kisswani, Telecommunications (interception and access) and its regulation in Arab countries. J. Int. Commer. Law Technol. (2010). https://doi.org/10.1504/ijlse.2011.041289
23. D.S. Griffith, The computer fraud and abuse act of 1986: a measured response to a growing problem. Vand. L. Rev. **43**, 453 (1990)
24. T. Curtiss, Computer fraud and abuse act enforcement: Cruel, unusual, and due for reform. Wash. Law Rev. (2016)
25. L.D. Roberts, D. Indermaur, C. Spiranovic, Fear of cyber-identity theft and related fraudulent activity. Psychiatry, Psychol. Law (2013). https://doi.org/10.1080/13218719.2012.672275
26. R. Richardson, M. North, Ransomware: evolution, mitigation and prevention. Int. Manag. Rev. (2017)
27. E. Ronen, A. Shamir, A.O. Weingarten, C. Oflynn, IoT goes nuclear: creating a Zigbee chain reaction. IEEE Secur. Priv. (2018). https://doi.org/10.1109/MSP.2018.1331033
28. M. Malik, Y. Singh, A review: DoS and DDoS attacks. Int. J. Comput. Sci. Mob. Comput. (2015)
29. S.G. Brandl, History of criminal investigation, In *Encyclopedia of Criminology and Criminal Justice* (2014)
30. History of Criminal Investigation, in *Handbook of Criminal Investigation* (2020)
31. Y. Prayudi, A. SN, Digital chain of custody: state of the art. Int. J. Comput. Appl. (2015). https://doi.org/10.5120/19971-1856
32. A. Venčkauskas, J. Toldinas, Š. Grigaliūnas, R. Damaševičius, V. Jusas, Suitability of the digital forensic tools for investigation of cyber crime in the Internet of Things and Services (2015). https://doi.org/10.18638/rcitd.2015.3.1.67
33. A. Ajijola, P. Zavarsky, R. Ruhl, A review and comparative evaluation of forensics guidelines of NIST SP 800-101 Rev.1:2014 and ISO/IEC 27037:2012 (2014). https://doi.org/10.1109/WorldCIS.2014.7028169
34. E. Casey, *Handbook of Digital Forensics and Investigation* (2010)
35. A. De Donno, *Handbook of Digital Forensics and Investigation* (2010)
36. J. Sremack, *Big Data Forensics—Learning Hadoop* (2015)
37. E. Hutchins, M. Cloppert, R. Amin, Intelligence-driven computer network defense informed by analysis of adversary campaigns and intrusion kill chains (2011)
38. S. Li, K.K.R. Choo, Q. Sun, W.J. Buchanan, J. Cao, IoT forensics: Amazon echo as a use case. IEEE Internet Things J. (2019). https://doi.org/10.1109/JIOT.2019.2906946
39. U.S. Food and Drug Administration, Cybersecurity Vulnerabilities Identified in St. Jude Medical's Implantable Cardiac Devices and Merlin Home Transmitter: FDA Safety Communication. U.S. Food and Drug Administration (2019)

40. F. Servida, E. Casey, IoT forensic challenges and opportunities for digital traces. Digit. Investig. (2019). https://doi.org/10.1016/j.diin.2019.01.012
41. D. Quick, K.K.R. Choo, IoT device forensics and data reduction. IEEE Access (2018). https://doi.org/10.1109/ACCESS.2018.2867466

Chapter 5
Supervised Deep Learning for Secure Internet of Things

This chapter elaborates on the potential of supervised deep learning solutions for fulfilling the security requirements of IoT-based systems with main aim to realize a reliable and trustworthy IoT environment. First, a detailed discussion of Convolutional Neural Networks is presented in terms of architecture, layers, and working methodology, and then the discussion go argue the advanced versions of CNN and its applications in IoT. Then, the chapter discuss the different forms of Recurrent Neural Networks (RNNs) in terms of architecture, calculations, problems, and improvements. Following this, the state-of-the-art attention network and Transformer networks is explored and investigated against standard networks. After that, the emerging graph networks is deliberated in supervised settings along with the their structures, advantages, applications, etc. Furthermore, the chapter discuss the state-of-the-art public datasets that can be used to. Finally, a fine-grained taxonomy is presented to categorize the supervised deep learning according to different perspectives of IoT community.

To sum up, this chapter intends to provide an overview Supervised Deep Learning approaches that could be exploited to develop a security solution in IoT environment. This is comprehensive discussed through the following sections:

- Convolutional Neural Networks
- Advanced Convolutional Networks
- Temporal Convolutional Networks
- Recurrent Neural Networks
- Graph Neural Networks
- Supervised Datasets and Evaluation Measures
- Taxonomy of Supervised Deep Learning
- Summary and Learnt Lessons.

Electronic Supplementary Material The online version of this chapter (https://doi.org/10.1007/978-3-030-89025-4_5) contains supplementary material, which is available to authorized users.

5.1 Convolutional Neural Network

Convolutional Neural Networks (CNNs) are considering a popular supervised deep learning model that witness a great success though the last years in different application domains. They are recognized as keystone for computer vision tasks such as segmentation, visual recognition, object detection, etc. Motivated by the success of CNNs, they come to be indispensable choice for endless IoT applications going from vision in self-driving vehicles, remote sensing and satellite intelligence. Although CNNs have been widely studies for many application domains, their potentials for being applied to secure Internet of Things (IoT) still in early phases.

5.1.1 Convolutional Layer

The convolution layer (CONV) employs a learning kerels to carry out convolution processes by scanning the input elements in regard to along its dimensions. The hyperparameters of CONV involve the size of kernel K, padding P, as well as stride S. The kernel K is the agent of dimension that surf the input to learn its representation. The stride S represent the number of input element by which the kernel shuffles following each operation. Zero-padding represent the method of inserting zero-elements to each side of the boundaries of the input. This value can be specified by hand or routinely set using any of the three methods known as "Valid", "Same", "Full". The output of convolution layer can be activate using any activation functions discussed in earlier chapters.

5.1.2 Pooling Layer

The pooling layer is popular layer commonly used for down-sampling procedure, habitually employed after a convolution layer, which does some spatial invariance. Specifically, max and average pooling are different types of pooling operation that seek to calculate the maximum and mean value of input, correspondingly.

5.1.3 Fully Connected Layer

The fully connected layer (FC) receives a flattened input wherever each input element is attached directly to all neurons. Generally, full connected layers are presented by the end of network to calculate the final decision i.e., class probabilities in classification problem.

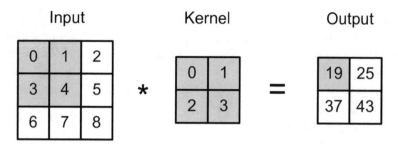

Fig. 5.1 Example of convolutional operation on two-dimensional input

5.1.4 Feature Map and Receptive Field

Generally, the output of convolutional layers is occasionally termed as a feature map, which could be interpreted as the learned or extracted features (representations) from the spatial facets of the current input. In forward propagation of CNNs, given element x belong to a particular convolutional layer, its receptive field describes all the features belonging in the current and earlier layers and contribute to the computation of the value x it worth mentioned that the size of receptive field might be greater than the real input size.

Let's interpret the meaning receptive field from the Fig. 5.1. Having convolution kernel 2×2, the receptive field of colored element in the output map (i.e., 19) is computed based on four colored elements in the input. Given a feature map M of size 2×2, let's contemplate a deeper CNN with one more 2×2 convolutional kernels that receive Y as an input and generate an output map O of single element. This way, the receptive field of O is based on all elements of M, whilst the receptive field of Y incorporates all input elements. Thus, building a deeper network is essential to make the elements of feature map have a greater receptive field to distinguish input features throughout a wider region.

5.2 Advanced Convolutional Networks

5.2.1 VGG Network

Inspired by the chip design domain, in which engineers turned from putting transistors to plausible components to logic blocks, the researchers improved the construction of CNN based on block design rather than individual neurons or layers. The notion of utilizing block-based design originally developed by the Visual Geometry Group (VGG) at Oxford University, which is termed as VGG network. specifically, the fundamental construction block of traditional CNNs is a series of the next (1) a padded convolution layer to retain the resolution, an activation function to introduce nonlinearity, a pooling operation. Single VGG block contains of a series of convolutional layers that have a padding of 1, attached to a max pooling layer for downsampling purpose [1].

5.2.2   Residual Network

Residual Network pursues the design VGG with convolution (3×3), where the residual block contains two convolution layers (3×3) with identical number of filters, each of them is accompanied with a batch normalization layer and a *ReLU* activation. Later, the input maps are added immediately to the maps prior to the last *ReLU* activation via skip connection (known as residual connection) without passed to beforementioned convolutional layers. This design scheme necessitates the size of input to be identical to the output of convolutional layers to be able to combine together. When the number of convolutional maps need to be changed, convolutional layer (1×1) can be added in residual connection to convert the input into the required shape for the extra procedure [1, 2].

5.2.3   Dense Network

Dense Network is common convolutional network that was introduced as an extension to the residual network. In particular, in dense block, the input of each convolutional layer is designated as concatenation of output of all previous convolutional layers and the input of the block. the major variations between ResNet and DenseNet stem from that dense block *concatenate the output of layers* outputs (represented as [,]) instead of added the input to the output in case of ResNet (Fig. 5.2).

The second difference is the large number of skip connections. Two key modules that constitute the DenseNet namely *dense blocks* and *transition layers*. The *dense blocks* describe the way of concatenating the inputs and outputs, whilst the *transition layers* regulate the number of channels so as to it is not overlarge [2, 3] (Fig. 5.3).

5.3   Temporal Convolutional Network

In the following, we will learn about temporal convolutional network and its fundamental architecture and building blocks. It is motivated by latest convolutional designs for chronological data and bring together minimalism, autoregressive modeling, and prolonged remembrance. The design of temporal convolutional network is motivated by two fundamental standards. First, the convolutions are causal, which means that there is no leakage of information from upcoming to

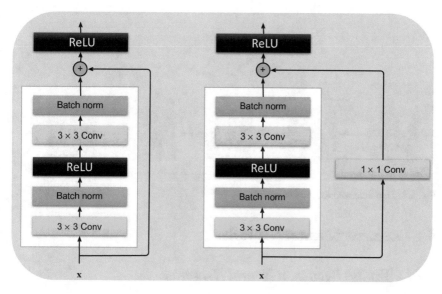

Fig. 5.2 The architecture of residual block in Residual Neural Network based on two-dimensional convolution

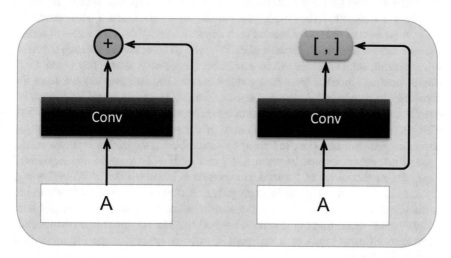

Fig. 5.3 The key difference between DenseNet (rightside) and ResNet (leftside)in connections

history. Secondly, the structural design enables receiving any sequential input and map it to targeted output format based on the underlying problem. Therefore, the temporal convolutional network employs causal convolutions, where the $t-$th output is convolved only with $t-$th element and the elements that precede them. The second standard is realized using the Temporal Convolution Network by setting the size of hidden layer identical as the input layer (Fig. 5.4).

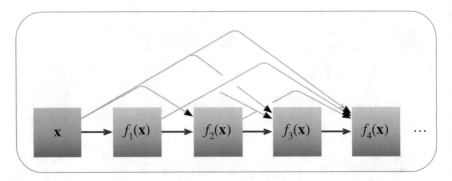

Fig. 5.4 Dense connections in DenseNet

5.4 Recurrent Neural Networks

5.4.1 Vanilla Recurrent Neural Networks

In a normal feedforward neural network (FFN), the inputs are almost autonomous. Nevertheless, with the sequential nature of IoT data, it is required to be aware about the historical input to better model the current input. A sequence here refers to a well-ordered set of data substances. For example, a sensory data is a series of sensor measurements. Assuming some application need to forecast the following sensor measurement; achieving this require remember the previous words. A standard FNN cannot precisely predict the sensory measurement because they do not have the capability to remember the earlier sensor data. To tackle this kind of limitation, deep learning research community introduced **recurrent neural networks (RNNs)**. This section provides a general overview of the role of RNNs for modeling the sequential data streams (i.e., IoT data) by remembering the earlier input elements. First of all, the difference between RNN and FNN is introduced and explained. Then, the methodology of forward propagation is inspected for RNN. Following, the training of RNN using the **backpropagation through time (BPTT)** algorithm is discussed. Later, we will look at the vanishing and exploding gradient problem, which occurs while training recurrent networks. After all, the discussion extends to address different kinds of RNN models, their design and usage for supervised learning (Fig. 5.5).

5.4.1.1 RNN Versus FNN

To understand the how the RNN differ from FNN, the Fig. 5.3 presents the illustrative representation for RNNs. It could be noted that the hidden layer of the RNN has a self-looped connection meaning the usage the previous hidden state in company with the current input element during the estimation of current output.

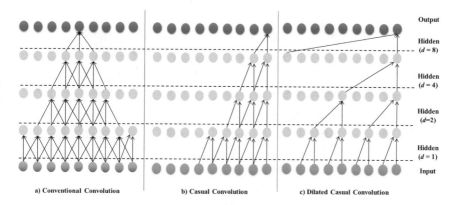

a) Conventional Convolution   b) Casual Convolution   c) Dilated Casual Convolution

Fig. 5.5 Comparison between conventional, casual, dilated casual convolution layers

Still mystified? Let us take a look at the unrolled representation of the RNN as shown in Fig. 5.3, in which the network is rolled out for a complete sequence. As illustrated, at the time step $t = 1$, the output is estimated according to the current input element x and the preceding hidden state a. Likewise, at time step $t = 2$, is the output predicted using the current input element and the earlier hidden state. This visual explanation of the working methodology of RNN can be mathematically formulated as follow:

$$h^{\langle t \rangle} = f1\left(W_{hh} \cdot h^{\langle t-1 \rangle} + W_{xh} \cdot x^{\langle t \rangle} + b_h\right) \qquad (5.1)$$

$$y^{\langle t \rangle} = f2\left(W_{hy} \cdot h^{\langle t \rangle} + b_y\right) \qquad (5.2)$$

where $f1, and f2$ represent the *tanh* activation function and *softmax* activation function, respectively. W_{xh} represent the input-to-hidden weight matrix. W_{hh} represent the hidden-to-hidden weight matrix. W_{hy} represent the hidden-to-output weight matrix. b_h and b_y represent the bias value of input and output state correspondingly. For more clarification, the previous calculations are visually represented as follow:

5.4.1.2 Backpropagation Through Time

Backpropagation Through Time (BPTT) refer to the application of the backpropagation phase during the training of RNN using sequential or temporal data. Once the forward propagation of the RNNs get completed the as above debated. Now, the backpropagation phase come into turn, in which particular loss is computed at each time step t, to estimate how well the network is predicting its output. For example, assuming that the cross-entropy loss is employed to train the RNN, then the network loss at a given time step can formulated as follows:

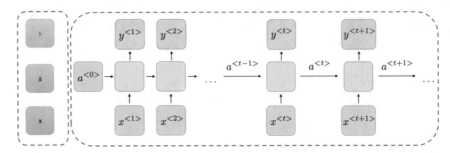

Fig. 5.6 Architecture of recurrent neural network

$$L^{\langle t \rangle} = -\overline{\overline{y}}^{\langle t \rangle} \log\left(y^{\langle t \rangle}\right) \tag{5.3}$$

where $\overline{\overline{y}}^{\langle t \rangle}$ represent the actual output as given in the labeled dataset, and $y^{\langle t \rangle}$ is the predicted output at a time step t. Hence, the final network loss can be computed as a sum of the losses as given as follows (Fig. 5.6):

$$L = \sum_{t=0}^{T-1} L^{\langle t \rangle} \tag{5.4}$$

Once the loss computed, the network seeks to minimize the loss value thought out training iterations this typically accomplished updating the network parameters (i.e., weights) of the RNN such as W_{ax}, W_{aa}, W_{ya}. Thus, to minimize this loss, the gradient descent algorithm employed to find the optimal value of these parameters by computing the gradients of the loss function (i.e., cross entropy) regarding to all the parameters; then, update the parameters according to the following update rule:

$$W_{xh} = W_{xh} - \alpha \frac{\partial L}{\partial W_{xh}} \tag{5.5}$$

$$W_{hh} = W_{hh} - \alpha \frac{\partial L}{\partial W_{hh}} \tag{5.6}$$

$$W_{hy} = W_{hy} - \alpha \frac{\partial L}{\partial W_{hy}} \tag{5.7}$$

To keep things simple, this section did not deeply discuss math behind the gradient calculation as it out of scope of this chapter. However, it can be beneficial for readers interested in an in-depth understanding for the BPTT applications in RNN.

5.4.1.3 Gradient Vanishing and Exploding Problems

According to working methodology of BPTT, the gradient of loss calculated with respect to all the weights in RNNs. However, this operation usually confronts with a problem known as the vanishing and exploding gradients. While computing the derivatives of loss with respect to W_{hh}, and W_{hy}, its cloud be noted that the backpropagation operation traverses all the road back to the initial hidden state, thanks to the dependency of $t - th$ hidden state $h^{(t)}$ on the hidden state at previous timestep $h^{(t-1)}$. For example, the gradient of loss $L^{(2)}$ regarding to W_{hh} is formulated as follow:

$$\frac{L^{(2)}}{\partial W_{hh}} = \frac{\partial L}{\partial y^{(2)}} \frac{\partial y^{(2)}}{\partial h^{(2)}} \frac{\partial h^{(2)}}{\partial W_{hh}} \tag{5.8}$$

Consider the term $\frac{\partial h^{(2)}}{\partial W_{hh}}$ in the above equation, the derivative of $h^{(2)}$ respecting W_{hh} cannot be directly computed because the $h^{(2)} = \tanh\left(W_{hh} \cdot h^{(1)} + W_{xh} \cdot x^{(2)} + b_h\right)$ is a function computed based on W_{hh} and $h^{(1)}$. Hence, the derivative regarding $h^{(1)}$ is also required. In the same way, $h^{(1)} = \tanh\left(W_{hh} \cdot h^{(0)} + W_{xh} \cdot x^{(1)} + b_h\right)$ denote a function that is reliant on W_{hh} and $h^{(0)}$. Thus, the derivative regarding $h^{(1)}$ is also required. As presented in Fig. 5.3, process of computing the $L^{(2)}$ derivatives traverse all the way back to the starting hidden state $h^{(0)}$, because each hidden state is reliant on its preceding hidden state.

To clearly understand the problem here, it worth noting that the backpropagation towards the first hidden state cause the gradient information gets progressively lost, thereby the RNN cannot backpropagate completely. In particular, at each time step t, the result from the calculation of the derivative $h^{(t)} = \tanh\left(W_{hh} \cdot h^{(t-1)} + W_{xh} \cdot x^{(t)} + b_h\right)$ during the back propagation is often in range from 0 and 1, hence, multiplication of such small values result in a smaller values. With small values of initial weights of RNN, the multiplication of derivatives and weights during the backpropagation basically entail a small number. Hence, the multiplication of those smaller values at every backward step makes the network gradient turn out to be infinitesimally tiny and resulting in a number impractical to handle by computer, this phenomenon is known as the **vanishing gradient problem**.

The vanishing gradients problem take place in different deep networks that use sigmoid or tanh as the activation function and not only in RNN. Thus, as typical solution of this problem can be the use of ReLU activation rather than tanh. In another way, an improved version of the RNN is introduced to effectively solve the vanishing gradient problem effectively, called the long short-term memory (LSTM) network, which is discussed in later section.

In opposite way, if the RNN is initialized with large value of weights, then the gradients will come to be extremely large at each time-step. During the backpropagation, the multiplications of these large numbers could result in an infinity that is impractical to handle by computer. This phenomenon is known as the gradient exploding problem.

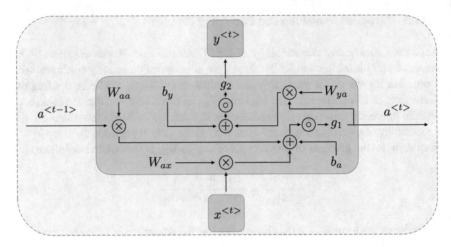

Fig. 5.7 Calculation of the parameters of recurrent neural network through forward propagation

In an attempt to address the gradient exploding problem, a gradient clipping method introduce to normalize the gradients based on a vector normalization (i.e., *L2*) and quote the gradient value to a specified range. For example, setting the clipping threshold to be 0.8 imply that the gradients should be kept in range from −0.8 to 0.8. when the value of the gradient surpasses −0.8, then it automatically changed to be −0.8, and likewise, when it surpasses 0.8, then it automatically changed to be 0.8. Given the loss L, then the gradient g regarding to W_{hh} is given as follow:

$$g = \frac{L}{\partial W_{hh}} \tag{5.9}$$

Then, the gradient is normalized with the L2 norm, i.e., g. When the normalized g go beyond the specified threshold, the gradient get updated as follows (Fig. 5.7):

$$g = \frac{threshold}{g} \cdot g \tag{5.10}$$

5.4.1.4 Different Categories of RNNs

According to the number of inputs and number of outputs, the RNNs can be divided into four main categories including one-to-one RNN, one-to-many RNN, many-to-one RNN, and many-to-many RNN we will look at a different type of RNN architecture that's based on numbers of input and output.

One-To-One RNN

In this category, one output is derived from one input, and the output from the time step *t* is passed as an input to the following time step. A typical example for that is the text generation task, in which the output generated from a present time step and feed it as the input to the subsequent time step to produce the following word. Another common usage of this category in stock market forecasts in IIoT system. For convenience and a clearer interpretation, the Fig. 5.5 shows the systematic architecture of the one-to-one RNN.

One-To-Many RNN

In a **one-to-many RNN**, multiple outputs and manifold hidden states is driven from one input. in other words, an output sequence can be generated from RNN using single input. Even Though a single input value available, the RNN share the hidden states through time domain to forecast the output. Different from above mentioned one-to-one RNN, the **one-to-many RNN** only share the earlier hidden states across time steps rather than earlier outputs. A typical application of this categories is the generation of image captions, wherein a single image fed as an input to RNN to generate as an output a sequence of words representing the image caption. As shown in the Fig. 5.6, a single image is passed as an input to the RNN, and at the first time step the word *Horse* is predicted; on the second time step,$t = 1$, the earlier hidden state is exploited to estimate the following word which is *standing*. Likewise, it lasts for a series of steps and forecasts the following word till the caption is produced.

Many-To-One RNN

In **many-to-one RNN,** as the name implies, one output value derived from sequence of input. One such popular example of a many-to-one architecture is anomaly detection. Wherein a sequence of industrial time series (i.e., from sensors, or machines) is fed as an input to RNN to compute the output at the end. As displayed in the Fig. 5.7, at each time step, a single sensory is fed in conjunction with the earlier hidden state as an input; and, at the last time step, it predicts whether the input time series is normal or abnormal.

Many-To-Many RNN

In **many-to-many RNN**, a sequence of output can be derived from sequence of input values. This input and output sequences can have different lengths i.e., language translation, or have an identical length as with audio generation. Given an English statement *"artificial intelligence broadly defined as human-like intelligence"* to be translated to Spanish, hence, this input would produce output the following output., *"inteligencia artificial ampliamente definida como inteligencia similar a la humana"*.

5.4.2 Longest Short-Term Memory (LSTM)

The main downside of **RNN** is that its memory could not keep hold of historical information for a long time. It is broadly known that an RNN stave the past sequential information in the hidden state, however as soon as the length of input become very long, it could not completely preserve the information in its memory because of the vanishing gradient trouble, as argued in the previous section. In an attempt to tackle this, an improved version of RNN, known as **long short-term memory (LSTM)** network, was presented to handle the gradient vanishing issue via specific composition known as a **gate**. Gates maintain the learnt information in memory for long enough time. Also, they learn to control the memory to determine which information to be maintained and which information to be discarded. In view of this, the following subsections discuss the architecture of LSTM, the way the LSTM resolved the deficiencies of vanilla RNN.

5.4.2.1 LSTM Architecture

Thanks to the vanishing gradient dilemma, the RNN can be trained inappropriately making the RNN unable to preserve prolonged sequences in the memory. For clarity, assume that input sequence is a short sentence *"The tree is __"*.

Then, it is easy for RNN to forecast the blank as *blue* according to the previously noticed information. On the other hand, given input sequence of long sentence as follow: *"Mohammed resided in Spain for 20 years. He likes playing football. He is a big-fan of He is a big-fan of Cristiano Ronaldo. He is fluent in ____."*

Then, it is easy for readers to predict the blank to be Spanish because they can remember the first sentence and knew that Mohamed spent 20 years living in Spain, which highly suggest that Mohammed is likely to be fluent in Spanish. Since RNN cannot preserve such amount of information in its hidden state, it cannot recall information for a lengthy time in its memory, leading to improper prediction of the blank word. This observed problem is not limited to NLP tasks but also extended to all tasks containing long sequences as an input. Hence, LSTM come into sight to alleviate these issues.

In this context, some question always arises when discussing the LSTM model. i.e., What makes LSTM so different? How do LSTM capture long-term dependency? How does it realize what information to maintain and what information to throw away from the memory?

The gating mechanisms in LSTM provide answer for all these questions. In an RNN, the hidden state employed for two objectives: first for saving the information and second for making estimates. Different from RNN, in the LSTM split the hidden states into two states, namely the **cell state** and the **hidden state**. The cell state is location of saving the sequential information, sometimes referred to as internal memory. The hidden state is employed to calculate the output estimates. The two states are shared within every time step. Conceivably, LSTM's building is encouraged by the computer's logic gates. LSTM introduces a *memory cell* (or *cell* for short) that has the same shape as the hidden state (some literatures consider the memory cell as a special type of the hidden state), engineered to record additional information. To manage the cell state, some gates are highly required. One gate is required to deliver the output from the cell, that is known as the *output gate*. Another gate is required to make a decision when to fed information to the cell, which is called the *input gate*. Besides, additional gate is required to reset the cell's content, which termed as forget *gate*. The motive behind this design is to empower the network to determine when to recall and when to disregard inputs in the hidden state through a devoted mechanism. In particular, these three gates are accountable for determining what information to add up, yield, and ignore from the memory making the LSTM more efficient for preserving the sequential or temporal information in the memory just as long as needed (Fig. 5.8).

Forget gate: The forget gate, $f^{(t)}$, take the responsibility of determining the information that have to be eliminated from the cell state. For example, given the following sentence as the input sequence "*Mohammed likes drawing. He is living in Cairo. Marcos, on the opposite, prefer going gym.*" Once the statement begins talking about Marcos, the network realizes that the subject has been changed from Mohammed to Marcos, and the information about Mohammed is likely to be no longer needed. Hence, the forget gate will get rid of information concerning Harry from the cell state. The operation of forget gate is dominated by a sigmoid activation. At each time step t, the input $x^{(t)}$ and the earlier hidden state $h^{(t)}$ are concatenated

Fig. 5.8 Architecture of one-to-one Recurrent neural network

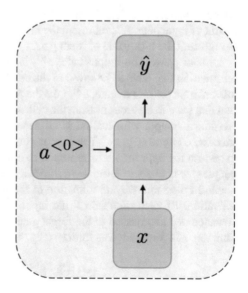

and passed to the forget gate. This function yield 0 when the specific information has to be eliminated from the cell state, and yield 1 when there is no information to be deleted. The forget operation in particular time step t is formulated as follows:

$$f^{\langle t \rangle} = \sigma \left(W_{hh}^{f} \cdot h^{\langle t-1 \rangle} + W_{xh}^{f} \cdot x^{\langle t \rangle} + b_f \right) \tag{5.11}$$

where σ denote the sigmoid activation, W_{xh}^{f} represent the input-to-hidden weight matrix of forget gate. W_{hh}^{f} represent the hidden-to-hidden weight matrix of forget gate. b_f represent the bias value of forget gate.

Input gate: The input gate, $i^{\langle t \rangle}$, take the responsibility of determining the information that have to be saved to the cell state. Consider the previous example with the input sequence "*Mohammed likes drawing. He is living in Cairo. Marcos, on the opposite, prefer going gym.*"

Following the elimination of forget gate from the cell state, the input gate determines the information to be kept in the memory. At This Juncture, since the information about *Mohammed* is deleted from the cell state, the input gate $i^{\langle t \rangle}$ take a decision to update the cell state with information regarding *Marcos*. In the same way, the function of forget gate is administered with a sigmoid activation. At each time step t, the input $x^{\langle t \rangle}$ and the earlier hidden state $h^{\langle t \rangle}$ are concatenated and passed to the forget gate. This function yield 0 when the specific information has to be eliminated from the cell state, and yield 1 when there is no information to be deleted. The input gate operation in particular time step t is formulated as follows:

$$i^{\langle t \rangle} = \sigma \left(W_{hh}^{i} \cdot h^{\langle t-1 \rangle} + W_{xh}^{i} \cdot x^{\langle t \rangle} + b_i \right) \tag{5.12}$$

where σ denote the sigmoid activation, W_{xh}^{i} represent the input-to-hidden weight matrix of input gate. W_{hh}^{i} represent the hidden-to-hidden weight matrix of input gate. b_i represent the bias value of input gate.

Output gate: The output gate, $o^{\langle t \rangle}$, take the responsibility of determining the information that have to be saved to the cell state. Since the cell state contain much information, the output gate, $o^{\langle t \rangle}$, take the responsibility of determining the information that have to be brought from the cell state to provide as an output. For example, consider example with textual input sequence "*Mohammed have achieved great success. Congrats_____.*" The output gate $o^{\langle t \rangle}$ search the cell state's information to choose the appropriate information to complete the sentence. At This Point, the output gate is likely to estimate or forecast *Mohammed* as appropriate choice for the blank. In the same way, the function of forget gate is administered with a sigmoid activation. At each time step t, the input $x^{\langle t \rangle}$ and the earlier hidden state $h^{\langle t \rangle}$ are concatenated and passed to the forget gate. The output gate operation in particular time step t is formulated as follows (Fig. 5.9):

$$o^{\langle t \rangle} = \sigma \left(W_{hh}^{o} \cdot h^{\langle t-1 \rangle} + W_{xh}^{o} \cdot x^{\langle t \rangle} + b_o \right) \tag{5.13}$$

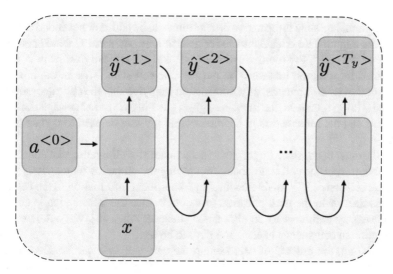

Fig. 5.9 Architecture of one-to-Many Recurrent neural network

where σ denote the sigmoid activation, W_{xh}^o represent the input-to-hidden weight matrix of output gate. W_{hh}^o represent the hidden-to-hidden weight matrix of output gate. b_o represent the bias value of output gate.

Cell State Update

Now, after recognizing the working methodologies of three gates of LSTM cell, iy is the time to understand how the cell state can actually updated by inserting pertinent new information and removing unnecessary information using those gates. First, to keep the new related information that should be inserted into the cell state, a new **internal state is calculated, which is** termed **candidate state** or $g^{(t)}$. Different from the gates controlled with, the candidate state is controlled by the tanh activation rather than sigmoid activation. The main reason behind that is the fact that the output of sigmoid function is always between 0and1, which mean forever positive. Also, it is highly required to have a function whose second derivative can withstand for a long span before turning to zero, so that, the network can avoid the gradient vanishment. It is required to permit the values of $g^{(t)}$ to be either positive or negative. Thus, the hyperbolic tangent function employed to returns values between -1 and $+1$. The candidate state operation in particular time step t is formulated as follows:

$$g^{(t)} = \tanh\left(W_{hh}^g \cdot h^{(t-1)} + W_{xh}^g \cdot x^{(t)} + b_g\right) \tag{5.14}$$

where W_{xh}^g represent the input-to-hidden weight matrix of candidate state. W_{hh}^g represent the hidden-to-hidden weight matrix of candidate state. b_g represent the bias value of candidate state. Thus, the candidate state retains all the new information to be injected into the cell state. Now, two essential questions arise. First, in what way the information in the candidate state is determined relevant or not? Second, In What

Way the candidate state information is determined to be added or not added to the cell state? Since input gate is accountable for determining whether to include new information or not, thus, the multiplication of $g^{(t)}$ and $i^{(t)}$ typically result in obtaining the related information that must be put into the cell state. That is, the input gate gives out 0 in case of unnecessary information implying that the cell state cannot be updated with $g^{(t)}$. Instead, the input gate gives out 1 in case of necessary information implying that the information in candidate state $g^{(t)}$ can be used to update the cell state.

On the other hand, since forget gate is accountable for eliminating the unnecessary information from the cell state, thus, multiplying the output of forget gate, $f^{(t)}$, and previous cell state $c^{(t-1)}$ enable retaining just pertinent information in the cell state. In particular, when the multiplication gives out 0, then the information in cell state is unnecessary and have to be eliminated. Instead, when it gives out 1, then the information in cell state is needed and have to be kept.

To sum up, the process of determining the new cell state can be accomplished using forget and input gates according to the following mathematical formula:

$$c^{(t)} = f^{(t)} \cdot c^{(t-1)} + i^{(t)} \cdot g^{(t)} \tag{5.15}$$

Hidden State Updates
After discussing the update of the cell state, the discussion continues elaborate how the information in LSTM's hidden state is updated (Fig. 5.10)

Since the output gate is accountable for determining the information to delivered as an output from the cell stat, hence, multiplication of the tanh of cell state $\tanh(c^{(t)})$, and outcome of output gate $o^{(t)}$, enable determining the new hidden state. Therefore,

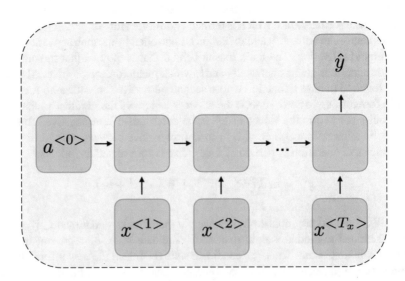

Fig. 5.10 Architecture of one-to-Many Recurrent neural network

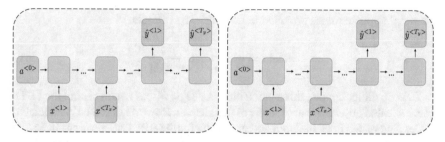

Fig. 5.11 Architecture of Many-to-Many Recurrent neural network with equal (right side) and non-equal (left side) mapping

the update of the hidden state can be computed according to the following formula:

$$h^{\langle t \rangle} = o^{\langle t \rangle} \cdot \tanh\left(c^{\langle t \rangle}\right) \tag{5.16}$$

5.4.2.2 Forward Propagation in LSTM

Laying it all together, the complete LSTM cell with all the functions is presented in Fig. 5.11. Where the hidden states and cell state are shared beyond time steps, implying that the LSTM calculates the hidden state $h^{\langle t \rangle}$ and cell state $c^{\langle t \rangle}$ at time step and transmits it to the following time step. The full forward propagation moves in the LSTM cell could be given as pointed bellow:

- Input gate calculation
- Forget gate calculation
- Output gate calculation
- Candidate state
- Cell state update
- Hidden state update
- Output state calculations

5.4.2.3 Backward Propagation in LSTM

Given LSTM with loss function L, to be minimized via gradient descent algorithm. Then, the derivative of loss L is computed with respect to the network weights and update the weights accordingly to minimalize the loss till reach the optimal weights in view of this, the LSTM has four inputs-to-hidden weight matrices W_{xh}^i, W_{xh}^f, W_{xh}^o, W_{xh}^g corresponding to the input gate, forget gate, output gate, and candidate state. It also has four hidden-to-hidden weight matrices W_{hh}^i, W_{hh}^f, W_{hh}^o, W_{hh}^g corresponding to the input gate, forget gate, output gate, and candidate state. We have one hidden-to-output weight matrix W_{hy}. the gradient decent update these weight matrices through the training to find the optimal weights, whereas the weight update rule is provided in the following formula:

$$\text{weight} = \text{weight} - \alpha \frac{\partial L}{\text{weight}} \qquad (5.17)$$

To keep things simple, this section did not deeply discuss math behind the LSTM gradient calculation as it out of scope of this chapter. However, it can be beneficial for readers interested in an in-depth understanding for the BPTT applications in LSTM. So, the interested reader can refer to [4].

5.4.3 Gated Recurrent Units

Since the LSTM provide a solution to gradient vanishing issue of the RNN using diverse set of gates as well as couple of states. However, it is worth noting that the LSTM architecture has large number of parameters because of the presence of multiple gates and two states. Therefore, during the backpropagation of the LSTM network, it is required to upgrade plenty of parameters within every single training iteration, thereby increasing the overall training time. In an attempt to solve this dilemma, Gated Recurrent Units (GRU) network was introduced as a simplification to the LSTM architecture. Different from the LSTM cell, the GRU cell contain just two gates namely reset gate and update gate and comprise single hidden state as well (see Fig. 5.12).

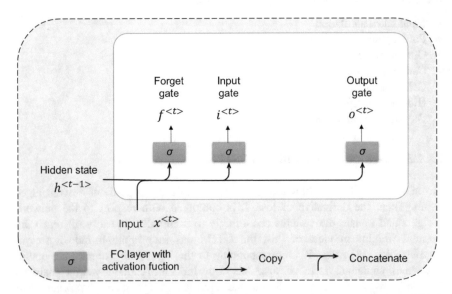

Fig. 5.12 Calculating the output of input gate, the forget gate, and the output gate in an LSTM Network

Update gate: the update gate takes the responsibility of determining which information to be kept from the previous time step, $h^{\langle t-1 \rangle}$, to be used in the calculation of the current time step $h^{\langle t \rangle}$. In other words, the update gate can be viewed as an integration of input gate and a forget gate previously discussed in in LSTM cells. Like LSTM gates, the update gate is controlled using sigmoid function as well. In any timestep t, the output of update gate, $z^{\langle t \rangle}$ can be defined as follows:

$$z^{\langle t \rangle} = \sigma\left(W_{hh}^{z} \cdot h^{\langle t-1 \rangle} + W_{xh}^{z} \cdot x^{\langle t \rangle} + b_z\right) \tag{5.18}$$

where σ denote the sigmoid activation, W_{xh}^{z} represent the input-to-hidden weight matrix of update gate. W_{hh}^{z} represent the hidden-to-hidden weight matrix of update gate. b_z represent the bias value of update gate.

Reset gate: The reset gate takes the responsibility of determining the way the new information is added to the memory, which mean amount of information to be forgotten. In any timestep t, the output of reset gate, $r^{\langle t \rangle}$ can be defined as follows:

$$r^{\langle t \rangle} = \sigma\left(W_{hh}^{r} \cdot h^{\langle t-1 \rangle} + W_{xh}^{r} \cdot x^{\langle t \rangle} + b_r\right) \tag{5.19}$$

where σ denote the sigmoid activation, W_{xh}^{r} represent the input-to-hidden weight matrix of reset gate. W_{hh}^{r} represent the hidden-to-hidden weight matrix of reset gate. b_r represent the bias value of reset gate.

Hidden State Updates: After understanding the role of reset and update gates, it is time to discuss the way these gates control the addition and/or removal of information for the purpose of updating the hidden state. First, the addition of new information to the hidden state is performed by creating a new state known as the content or candidate state, $c^{\langle t \rangle}$ to maintain the needed information. Since the reset gate is accountable for eliminating unnecessary information. Hence, it is employed to establish a content state, $c^{\langle t \rangle}$, that maintains merely the necessary information. In any timestep t, the content state, $c^{\langle t \rangle}$, can be defined as follows:

$$c^{\langle t \rangle} = \tanh\left(W_{hh}^{c} \cdot \left(h^{\langle t-1 \rangle} \cdot r^{\langle t \rangle}\right) + W_{xh}^{c} \cdot x^{\langle t \rangle}\right) \tag{5.20}$$

A nonlinearity introduced in form of $\tanh$ to guarantee that the values in the content and hidden state stay in the range $(-1, 1)$. Second, the outcome of the update gate to determine the contribution of previous hidden state $h^{\langle t-1 \rangle}$ and candidate state $c^{\langle t \rangle}$ to the new hidden state $h^{\langle t \rangle}$. Since the update gate, $z^{\langle t \rangle}$, is accountable for determining the information to be kept from the preceding time step, $h^{\langle t-1 \rangle}$, hence, the multiplication of $z^{\langle t \rangle}$ and $h^{\langle t-1 \rangle}$ can provide only the required information from the earlier step. Rather than establishing a new gate to calculate the contribution of candidate state $c^{\langle t \rangle}$ to the new hidden state, the complement of update gate is calculated as $1 - z^{\langle t \rangle}$, then multiplied with candidate state $c^{\langle t \rangle}$. Therefore, the new hidden state can be basically by calculating the elementwise convex combinations between $h^{\langle t-1 \rangle}$ and $c^{\langle t \rangle}$, as expressed below:

$$h^{\langle t \rangle} = \left(1 - z^{\langle t \rangle}\right) \cdot c^{\langle t \rangle} + z^{\langle t \rangle} \cdot h^{\langle t-1 \rangle} \tag{5.21}$$

where $\odot$ represent Hadamard product. When the value of $z^{\langle t \rangle}$ is nearly 1, then the update gate just holds the information at the old state. This typically imply that the current input information $x^{\langle t \rangle}$ is disregarded, successfully hopping timestep t in the interdependency sequence. On the other hand, When the value of $z^{\langle t \rangle}$ is nearly 0, the new hidden state $h^{\langle t \rangle}$ bear up the content state $c^{\langle t \rangle}$. This structural design enables addressing the gradient vanishing issue present in RNNs and better learn sequential dependencies over big time step spaces.

5.5 Graph Neural Networks

Graph Neural Networks (GNNs) can be defined as an improved deep learning algorithms that enable representation learning form graph-regulated data. The traditional deep learning models cannot be clearly simplified to graph-regulated data because the graph composition might not always be a regular grid. As a result of the great achievement made by deep learning in various discipline, academia and industry began to devote more efforts to the graph learning [5]. In this regard GNN can be categorized into node-centered and graph-centered representation learning. In node-centered scenarios, GNNs aim to learn insightful representations for each node in such a way that empower the node-centered tasks. In graph-centered scenarios, the GNNS seek to learn demonstrative features for the whole graph, whereas node-level representation learning is generally considered as a midway phase. The practice of node-level representation learning often make use of the features of input node as well as the graph construction [6]. Different type of graph is presented in the literature which can be summarized in the following figure.

5.6 Supervised Datasets and Evaluation Measures

5.6.1 Datasets

This sub-section offers an overview of the open-source datasets employed for developing supervised deep learning solution for IoT security.

5.6.1.1 KDD Cup99 Dataset

In 1998, an early work [7] was made by Defense Advanced Research Project Agency (DARPA) to create the Knowledge Discovery and Data Mining (KDD) dataset. it considered one of the most broadly used dataset for intrusion detection and the

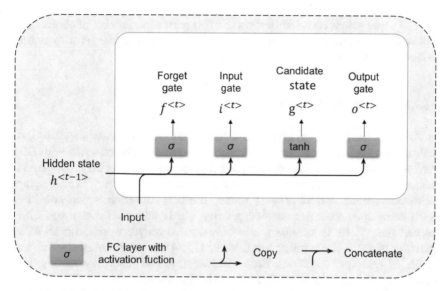

Fig. 5.13 Calculating the candidate memory cell in an LSTM model

most frequently used in the research literature. The data observations were gathered employing numerous internet-connected computers to model a small United State Air Force base of limited personnel, in which host log files and network packets were monitored and aggregated. The accumulated network packets comprising around 4,900,000 records. The test set of two weeks containing around two million of connection observations, each constituted with 41 features and is labeled as normal or abnormal. The abnormal observations divided into four categories of attacks namely denial of service (DoS), probe, remote to local (R2L), and user to root (U2R). Unfortunately, the efficiency and ability of this dataset to contemplate real-world environments have been extensively criticized for several shortcomings i.e., redundant observations, simulation artifacts, and the fact that it does not simulate the contemporary IoT traffic and attacks (Fig. 5.13).

5.6.1.2   NSL-KDD Dataset

In 2009, NSL-KDD is released by [8] as an improvement to the KDD dataset to solve its common deficiencies. In the KDD dataset, classification was biased on the subject of more repetitive observations. Nevertheless, in the NSL-KDD dataset, replicated observations were eliminated from training and testing data to stop the learning classifiers from realizing irrationally elevated recognition rates thanks to redundant observations. The dataset comprises four categories of attacks namely DoS, Probe, U2R, and R2L. There are 4,898,431 observations constitute the training set and

311,027 observations constitute testing set. This reasonable number of observations makes the dataset affordable to perform experiments on the entire set without need to random sampling.

5.6.1.3   UNSW-NB15 Dataset

In 2015, the authors of [9] presented UNSW-NB15 dataset containing 2,540,044 realistic observations of normal and abnormal network traffic captured with IXIA PerfectStorm tool exploiting three virtual servers, one of them was configured to produce the abnormal network traffic while the other two servers were configured to disseminate the normal network traffic. The data collection is performed in a small network environment containing 45 unique IP addresses over a small-time interval (i.e., 31 h). it includes a mix of real usual activities and artificial attack activities of the network traffic, resultant in 175,341 training observations as well as 82,332 testing observations. Each observation has 49 features involving basic features, packet features mined from the packet header and its payload, time or flow features engendered via the streaming of packets from a source to a destination, and extra generated features related to the statistical attributes of connections acquired from the raw network packets by means of Bro-IDS and Argus tools. The IXIA tool simulated nine categories of attacks including Attacks are categorized as Analysis, Backdoor, DoS, Exploits, Fuzzers, Generic, Reconnaissance. The major trouble with this dataset is the synthetic generation of traffic, which is related to hypothetical rather than realistic behaviors in the Internet.

5.6.1.4   DDoS 2016 Dataset

In 2016 [10], The dataset represents the simulated data gathered via Network Simulator NS2 in a regulated environment, which has four malevolent categories of network attacks namely Smurf, DDOS Using SQL injection (SIDDOS), UDP Flood, and HTTP Flood. The dataset contains 734,627 observations, each consist of 27 features and belong to one of five classes i.e., normal traffic class and the above-mentioned attack class.

5.6.1.5   UGR'16 Dataset

By 2016, the authors of [11] created UGR'16 Dataset in the University of Granada with main aim to evaluating the network intrusion detection systems. It was constructed with actual traffic and up-to-date attacks, where the data obtained from multiple netflow v9 collectors tactically situated in the network of a Spanish tier-3 Internet Service Provider (ISP). The data contain 16,900 million one directional flows acquired over four months. The full dataset encompasses two distinct subsets namely a *calibration set* and a *test set*. The real network traffic in the test set was combined with synthetically generated malicious attack flows caught in a regulated network environment that slightly reduces the quality of the dataset. The data flows are labeled either attack, or normal, where background imply that it is not identified whether the flow or packet comprises a malicious traffic.

5.6.1.6   CICIDS 2017 Dataset

In 2016, the authors of [12] build a ICIDS 2017 dataset with main priority to generate a realistic background traffic by abstracting the behavior of human interactions via B-Profile system. A thorough network structure was configured for data collection including Firewall, Modems, Routers, Switches, and nodes hosting a various operating system i.e., Apple's macOS iOS, Windows 7, 8.1, 10, and Linux. The dataset is totally annotated and has over 80 network traffic features extracted for each Netflow observation and evaluated for all benign and invasive flows using CICFlowMeter software which is freely obtainable at the Canadian Institute for Cyber Security website [13]. The data contain benign traffic samples and more than twelve attacks categories including Botnet, DDoS, DoS GoldenEye, DoS Hulk, DoS Slowhttptest, DoS slowloris, Heartbleed, FTP-Patator, PortScan, SSH-Patator, Web Attack Brute Force, Infiltration, and Web Attack XSS. The data is reported to have more than 2,800,000 observations [14].

5.6.1.7   CSE-CIC-IDS2018 Dataset

In 2018, the authors of [15] introduced CSE-CIC-IDS2018 Dataset, in which the notion of profiles (i.e., B-profiles, and M-Profiles) are employed to generate data in a systematic way, which will encompass exhaustive descriptions of intrusions and abstract distribution models for protocols, applications, or low-level network entities. These profiles applied to generate wide range of events on divers set of the network topologies. The full dataset includes seven distinct attack categories namely, network infiltration, Heartbleed, DoS, Botnet, Web attacks, DDoS, and Brute-force. The attacking infrastructure comprises fifty machines, and the victim enterprise involve five departments and contains 30 servers and 420 machines. The dataset incorporates the captures network traffic and system logs of each machine, accompanied by more than 80 features extracted from the acquired traffic utilizing CICFlowMeter-V3. The data was reported to have more than 16,000,000 observations [16].

5.6.1.8   IoT Network Intrusion Dataset

In 2019, The IoT Network Intrusion Dataset [17] was released and created using two typical smart home devices, a EZVIZ Wi-Fi Camera (C2C Mini O Plus 1080P), SKT NUGU (NU 100), set of few smartphones, and set of laptops. All of the involved devices were connected to one wireless network on which various attacks were mimicked utilizing tools such as Nmap. This dataset contains 42 raw network packet files acquired by utilizing watch mode of network adapter at various moments. Beside normal traffic, the dataset contains four main categories of attacks including Scanning attacks (Host Discovery, Port Scanning, OS/Version Detection), MITM, DoS, and Mirai botnet (i.e., Host Discovery, Telnet Bruteforce, UDP Flooding, ACK Flooding, HTTP Flooding). All attacks apart from Mirai Botnet are captured from packets imitating attacks via Nmap. On the other hand, the attacks of Mirai Botnet were spawned on a laptop and later engineered to make it seem as derived from the IoT device. The data was reported to have more than 2,985,994 observations.

5.6.1.9 N-BaIoT Dataset

In 2018, an effort introduced the N-BaIoT dataset [18], which was exclusively intended for network-based botnet recognition. The traffic data were collected from nine IoT devices connected many access points, switches, and routers over wired and wireless connections. The data contain two categories of botnet attacks namely BASHLITE (i.e., Scanning, Junk, UDP flooding, TCP flooding, COMBO) and Mirai (i.e., Scanning, Ack flooding, Syn flooding, UDP flooding, UDP flooding). Hence, each data observation could be belonging to one of ten classes of attack or benign class. Different from the IoT Intrusion Dataset, the features in this dataset are statistical measures resulting from sensor logs collected from each of the devices in the empirical set-up. Given the designers' emphasis on profiling device behavior, the existing data is individually supplied for each device. The data was reported to have more than 16,000,000 observations [16]. The data was reported to have 7,062,606 instances, each consist of 115 traffic features.

5.6.1.10 Aposemat IoT-23

A recent effort dedicated for developing network traffic flows from IoT devices, called Aposemat IoT-23 [19]. Its aim is to present a huge dataset of realistic and annotated IoT malware viruses and IoT benign traffic for research community. Aposemat IoT-23's IoT traffic is acquired from genuine physical IoT devices. The aggregated data within the Stratosphere project and contains three captures of benign IoT devices traffic and twenty captures of malware implemented in IoT devices. The dataset includes more than 760 million packets and 325 million annotated flows of more than 500 h of traffic captured taken during 2018 and 2019 at Czech Technical University. The malicious captures were designed by implementing a particular malware in a Raspberry Pi. The executed attacks comprise various botnets like Mirai, Hajime, Torii, Hide and Seek, etc. Above And Beyond the initial packet capture files, netflows generated with Zeek/Bro IDS and the corresponding labels are offered as well.

5.6.1.11 ToN_IoT Dataset

In 2020, ToN_IoT dataset was introduced by [20], which contain data from heterogeneous sources gathered from Telemetry datasets of IoT and IIoT sensors, heterogenous operating systems data i.e., Ubuntu 14, Windows 7, Windows 10, and Ubuntu 18 TLS, and Network traffic datasets. The data was aggregated from a practical and huge-scale testbed network constructed at the IoT Lab of UNSW Canberra Cyber, which connect multiple physical systems, virtual machines, hacking platforms, cloud, and fog platforms, and IoT sensors to simulate the scalable and composite nature of IIoT environments. The data is reported to consist of 22 million of log entries each with 43 features and two labels. The authors of [21] described the threat model used to mimic practical attacks for this dataset. These attacks are then scrutinized, and data elements are annotated as either normal or attack, along with the kind of attack is executed. The dataset entails nine categories of attacks including Scanning attack, Denial of Service (DoS) attack, Distributed Denial of Service (DDoS) attack, Ransomware attack, Backdoor attack, Injection attack, Cross-site Scripting (XSS) attack, Password attack, and Man-In-The-Middle (MITM) attack.

5.6.1.12 LITNET-2020 Dataset

In 2020, the authors of [22] published The LITNET-2020 dataset, which contain real-world network traffic data caught in the nfcapd binary format files. Whereas each nfcapd files have all NetFlow features, stretched with 19 custom attack related features resulting in a total of 85 network flow features. The network environment employs five exporters to monitor and acquire network flows. The data contain twelve class of attacks including Packet fragmentation attack, Spam bot's detection, Reaper Worm, Scanning/Spread, Internet Control Message Protocol (ICMP)-flood, HTTP-flood, Blaster Worm, LAND attack, Smurf, User Datagram Protocol (UDP)-flood, Code Red Worm, and Transmission Control Protocol (TCP) SYN-flood.

5.6.1.13 InSDN Dataset

In 2020, an effort was made to generate an attack-related dataset [23], which was completely aimed at intrusion detection Software-Defined Network (SDN). The data consists of three main parties depending on the types of traffic and the target machines. The former one only comprises normal traffic. The next party includes the attack traffics pursuing Mealsplotable-2 server. The third group contain attacks on the OVS machine are considered. The Tcpdump tool was employed to apprehend the traffic traces for each type at the SDN controller interface and objective machine, while the CICFlowMeter tool was employed to extract the flow features from this dataset. The data is reported to have 56 kinds of samples with more than 80 features which are divided into ten groups of features namely Network identifiers, Packet-based, Bytes-based, Interarrival time, Flow timers', Flag, Flow descriptors, and Subflow descriptors. The final data is reported to have a total 343,939 samples for normal and attack traffic, wherein the attack traffic contain 11 classes of attacks. these attacks could be initiated from inside or outside of the SDN network.

5.6.1.14   Novel SDN Dataset

In 2020, an effort was presented to generate an attack-related dataset [24], that completely target intrusion detection in SDN-enabled IoT environments. In the testbed setup, the Mininet Tool was employed for simulating the IoT network that was managed by an ONOS controller. It contains two datasets, where in the first, five hosts are employed as a simulated IoT devices sending insignificant amounts of data to the server regularly, while in the second, ten IoT devices are simulated. The data is reported to have 5 kinds of attacks (i.e., DoS, DDoS, Port scanning, OS fingerprinting, Fuzzing) plus one normal class, while the first dataset contain 27.9 million observations and the second dataset contain 30.2 million observations. Each observation was consisting of 33 features.

5.6.2   Evaluation Metrics

The deep learning algorithms introduced for IoT security (attack detection) often make use of following measures in the evaluation purposes, namely accuracy (A), precision (p), Recall (R), $F1$-score, True Positive Rate (TPR), False Alarm Rate (FAR), False Positive Rate (FPR), Receiver Operating Characteristic (ROC) curve, Area Under the Curve (AUC), and. Most of these performance measures can be calculated from a confusion matrix, which is a matrix-like representation of the classification outcomes. Whereas the True Positive (TP) and True Negative (TN) represent the number of attack and normal samples correctly categorized by the model. Meanwhile, False Positive (FP) and False Negative (FN) denote the number of normal and attack records erroneously categorized by the model. The structure of confusion matrix in attack detection problem is presented in Fig. 5.15.

Accuracy: it represents the proportion of correctly categorized IoT samples by the model to the total number of IoT samples. It is commonly believed that high accuracy of model is best. Right, accuracy is a good metric but only in case of symmetric datasets in which the number of false positives and false negatives are roughly same. The accuracy is calculated as follow:

$$\text{Accuracy}(A) = \frac{TP + TN}{TP + TN + FP + FN} \tag{5.22}$$

Precision: *it represents* the proportion of samples correctly categorized by the model to the total number of predicted samples. The accuracy is calculated as follow:

$$\text{Percision}(P) = \frac{TP}{TP + FP} \tag{5.23}$$

Recall or Sensitivity: it represents the proportion of samples correctly categorized by the model to the total number of actual class samples.

$$\text{Recall}(R) = \frac{TP}{TP + FN} \tag{5.24}$$

F1-score: it represents the subjective average, or the harmonic mean of the precision and recall of deep learning classifier. Thus, it takes into consideration the false positives as well as false negatives. Instinctively it is not as simple to comprehend like accuracy, however F1-score is typically more valuable than accuracy, mainly when the data undergo an imbalanced class distribution. Accuracy performs better when false negatives as well as false positives have comparable cost. Instead, when the costs of these two categories are highly diverse, it is favored to take F1-score into considerations. The f1-score is calculated as follow:

$$F1 - \text{score} = 2 \times \frac{P \times R}{P + R} = \frac{2TP}{2TP + FP + FN} \tag{5.25}$$

There are circumstances however for which it is required to assign a ratio more significance to either recall or precision. Which necessitate modifying the above formula a little such that it can incorporate a variable parameter beta β to fulfill this objective. This F1-score can be generalized for numerous classification dilemmas using beta-version $F1 - \text{score}_\beta$, which is computed as follow:

$$F1 - \text{score}_\beta = \left(1 - \beta^2\right) \times \frac{P \times R}{\left(\beta^2 \times P\right) + R} \tag{5.26}$$

where β parameter permits controlling the tradeoff of significance between recall and precision. More focus dedicated to precision when $\beta < 1$ while more focus dedicated for recall when $\beta > 1$.

True Positive Rate (TPR): it represents the proportion of samples correctly identified belonging to given class to the total number of samples in this class. The TPR is calculated as follow:

$$TPR = \frac{TP}{TP + FN} \tag{5.27}$$

False Positive Rate (FPR): it represents the proportion of samples incorrectly identified belonging to given class to the total number of samples in this class. The FPR is calculated as follow:

$$FPR = \frac{FP}{TN + FP} \tag{5.28}$$

Specificity: it represents the proportion of items correctly categorized not belonging to a given class to total number of samples in this class. It is correlated with FPR as follows (Fig. 5.19):

$$\text{Specificity} = 1 - \text{FPR} = \frac{\text{TN}}{\text{TN} + \text{FP}} \qquad (5.29)$$

The *receiving operating characteristics (ROC)* curve is the calculations of plotting TPR against FPR. Every point in the ROC space communicates to the performance of the deep learning model on a particular distribution. Hence, the ROC curve delivers an illustrative representation of the trade-off between the gains (TPR) and costs (FPR) of classification with regard to data distributions. In addition, the area under the ROC curve (AUC) is a relevant performance measure that evaluates the scale of separability that facilitates determining whether the deep learning model can successfully differentiate between different attack classes. Figure 5.16 illustrates the relationship between these two performance measures.

5.7 Taxonomy of Deep Learning Solutions for IoT

Many research efforts have been dedicated for solving broad range of security issues in IoT environment. This section aims to provide a fine-grained taxonomy for categorizing the cutting-edge deep learning solutions for IoT security based on different set of design facets including input scheme, detection scheme, deployment scheme, and evaluation scheme. Each facet is further divided into multiple categories, the proposed taxonomy is visually illustrated by Fig. 5.17.

5.7.1 Input Scheme

IoT data is considered the most essential part for developing and supervised deep learning solutions. The data samples could be aggregated from various sources, involving users' devices, network traffic, host logs, and application/service-related entities. This input data is labeled and employed to train deep learning model and can be produced in a real environment or simulated environment or mix of them. Nonetheless, the majority of the current studies use the present benchmark datasets to assess the deep learning solution in off-line mode, hence, they unqualified to support real-time security applications. Various methods have been developed to tackle the high dimensionality issue, like incessant traits discretization and feature learning.

The feature learning is a preparation that seek to engender a low-dimensional representation that can be effectively exploited to train a supervised model. This preparation method can be roughly categorized into deterministic probabilistic approaches. These approaches can be further categorized into linear and non-linear approaches (Fig. 5.14).

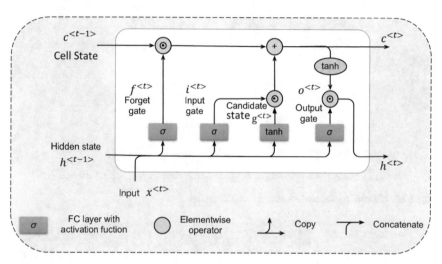

Fig. 5.14 Computing the hidden state in an LSTM model

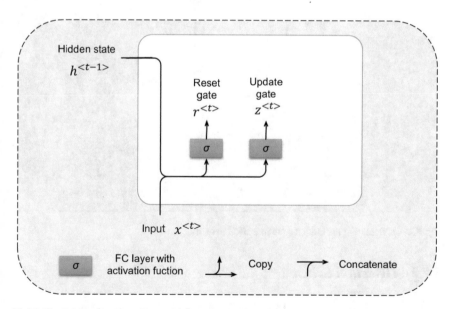

Fig. 5.15 Calculating the output of reset gate and the update gate in a GRU network

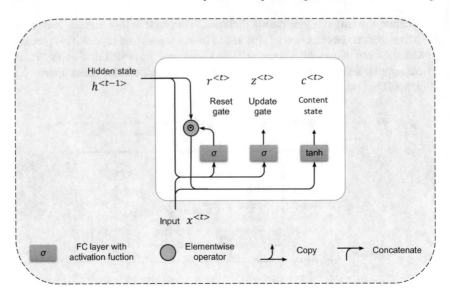

Fig. 5.16 Computing the content state in a GRU model

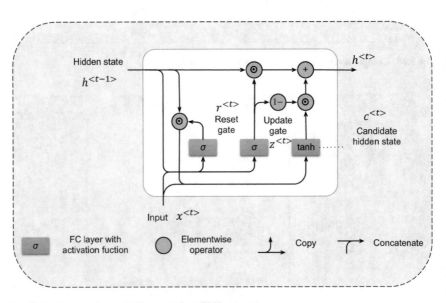

Fig. 5.17 Calculating the hidden state in a GRU network

5.7.2 Detection Scheme

Variety of supervised deep learning-based security solutions have been introduced, which are categorized based on network design into four main categories

including Convolutional Networks, Recurrent Networks, Graph Network, and Hybrid Networks. From output perspective, the security solutions based on deep learning classifier can be binary when the objective is to differentiate between regular and irregular behaviors and can be multi-class as soon as the intrusion it aims to target a particular attack category. Furthermore, from time viewpoint, the supervised solution can operate in either offline or online mode. The online mode makes the solution adequate for time sensitive IoT applications (Fig. 5.18).

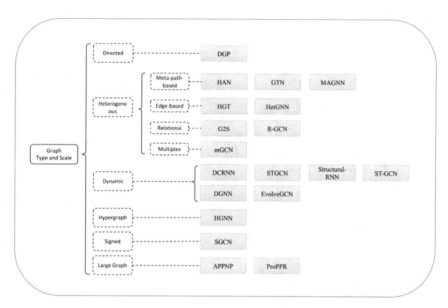

Fig. 5.18 A taxonomy of variations contemplating graph type and scale

Fig. 5.19 Structure of confusion matrix and its content

		Actual	
		Normal	**Attack**
Predicted	**Normal**	True Positive (TP)	False Negative (FN)
	Attack	False Positive (FP)	True Negative (TN)

5.7.3 Deployment Scheme

The deployment scheme decides the configuration and setup of the deep learning solutions, which could be categorized based on the nature of architecture to be centralized, distributed, or hierarchal. centralized configurations aggregate the data into central location where the training is taken place. In distributed configuration, the data and deep learning model are propagated between various physical locations. In hierarchal configurations, the data and process are moved up across IoT layers (Figs. 5.20 and 5.21).

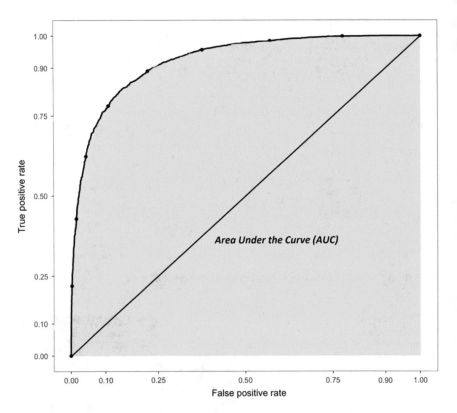

Fig. 5.20 Visualization of receiving operating characteristics over axis coordinate

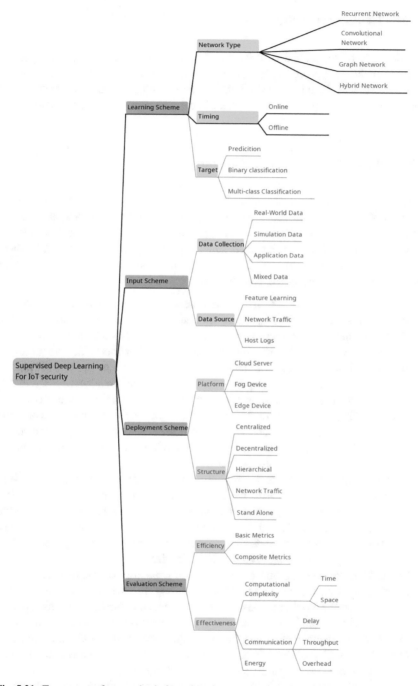

Fig. 5.21 Taxonomy of supervised deep learning approaches for supervised deep learning approaches for IoT security

5.7.4 Evaluation Scheme

The designer of supervised security solution can also categorize according to the target of evaluations process. Most of the present approaches emphasis on the detection performance based on evaluation measures. Other approaches emphasis on computational efficiency. Another target can be the communication efficiency.

5.8 Summary and Learnt Lessons

This chapter provides an in-depth discussion about the role of the supervised deep learning for securing IoT environments by enabling efficient and early detection of different attacks. Besides, it also provides a taxonomy for classifying the current supervised deep learning according to different perspectives. In the nutshell, this chapter delivers useful insights regarding the potential of deep learning for maintaining the security of IoT environments:

- With respect security of IoT infrastructure, the majority of efforts for attack/intrusion detection use the simulation datasets for experimentation. However, they do not thoroughly reflect the current complex and heterogeneous nature of real IoT environments, but they are beneficial for comparatively analyzing the performance of supervised deep learning models. It is difficult to realize a fair comparative analysis of all the supervised models addressed in this chapter owing to several reasons. First and leading, the research or industry efforts often use different datasets or different portions of the same dataset. Second, the variation in the evaluation metrics employed for evaluating the suggested supervised solution. Third, the variations in the applied data preparation. Forth, the experimental environment always varies from one to another.
- As a general noticeable trend, the performance of almost all supervised deep learning solutions outperforms the conventional and machine learning solutions even when contemplating noisy real-world IoT data. Moreover, preparing the training data often lead to higher performance. Likewise, eliminating unnecessary features enables reducing data dimensions and thereby improving computation efficiency while preserving detection performance.
- The exhaustive investigation reveals that the capability of the supervised deep learning solution to detect attacks/intrusions highly relies on the nature of the attack and on the number of classes taken into account during the training phase. Another critical conclusion is that the more a supervised solution is trained with the up-to-date features of attacks, the better the detection of popular and unknown (zero-day) attacks will be. Therefore, supervised detection solutions ought to be regularly trained with modernized IoT traffic samples.
- With respect security of IoT software, supervised deep learning solutions always outperform machine learning for malware, botnet, and ransomware detection.

- DL-based architectures are tested with different datasets, in this case even with a greater number of variants than in the infrastructure area. Furthermore, as obtaining reasonably malicious data is problematic, the proportion of normal to malevolent data in the training dataset is unbalanced which negatively influences the efficiency of supervised solution. Consequently, standardized balanced datasets are critical for forthcoming enhancements on supervised solutions to be able to be generalized to detect new malware.
- Supervised deep learning for malware detection fall in two core categories reliant on the utilization of API calls or opcodes. Android operating system is the major emphasis of malware detection in IoT environment as almost all malware targets are Android devices.
- The supervised detection of botnet and ransomware necessitate analyzing the network traffic to generalize ordinary features of botnets all along their life cycle. Unlike conventional intrusion detection techniques, supervised deep learning solutions can identify ransomware with interactive transformations and ransomware that utilizes packers to conceal. Besides, they find out the threat when the contamination process begins.
- Privacy can be considered as a key security concern in IoT environments which is an infancy field of cyber security requiring extra investigation. Privacy preservation efforts primarily follow three distinct methodologies: collaborative supervised deep learning, differential privacy, and training on encrypted data. These methodologies realize satisfactory degrees of performance. Nevertheless, they still subject to complexity and efficiency privacy trade-offs.

References

1. K. He, X. Zhang, S. Ren, J. Sun, "Deep residual learning for image recognition," (2016). https://doi.org/10.1109/CVPR.2016.90
2. Z. Li, F. Liu, W. Yang, S. Peng, J. Zhou, A survey of convolutional neural networks: analysis, applications, and prospects. IEEE Trans. Neural Networks Learn. Syst. (2021). https://doi.org/10.1109/tnnls.2021.3084827
3. M.Z. Alom, M. Hasan, C. Yakopcic, T.M. Taha, V.K. Asari, Inception recurrent convolutional neural network for object recognition. Mach. Vis. Appl. (2021). https://doi.org/10.1007/s00138-020-01157-3
4. S. Ravichandiran, *Hands-on Deep Learning Algorithms with Python: Master Deep Learning Algorithms with Extensive Math by Implementing them Using Tensorflow*. Packt Publishing Ltd (2019)
5. W. Cao, Z. Yan, Z. He, Z. He, A comprehensive survey on geometric deep learning. IEEE Access (2020). https://doi.org/10.1109/ACCESS.2020.2975067
6. Z. Zhang, P. Cui, W. Zhu, Deep learning on graphs: a survey. IEEE Trans. Knowl. Data Eng. (2020). https://doi.org/10.1109/tkde.2020.2981333
7. UCI Machine Learning Repository, "KDD Cup 1999 Data," *1999]*. https://Kdd.Ics.Uci.Edu/Databases/Kddcup99/Kddcup99.Html (2015)
8. M. Tavallaee, E. Bagheri, W. Lu, A. A. Ghorbani, "A detailed analysis of the KDD CUP 99 data set," (2009). https://doi.org/10.1109/CISDA.2009.5356528

9. N. Moustafa, J. Slay, "UNSW-NB15: A comprehensive data set for network intrusion detection systems (UNSW-NB15 network data set)," (2015). https://doi.org/10.1109/MilCIS.2015.734 8942
10. M. Alkasassbeh, G. Al-Naymat, A. B.A, M. Almseidin, "Detecting distributed denial of service attacks using data mining techniques." Int. J. Adv. Comput. Sci. Appl. (2016). https://doi.org/ 10.14569/ijacsa.2016.070159
11. G. Maciá-Fernández, J. Camacho, R. Magán-Carrión, P. García-Teodoro, R. Therón, UGR'16: A new dataset for the evaluation of cyclostationarity-based network IDSs. Comput. Secur. (2018). https://doi.org/10.1016/j.cose.2017.11.004
12. I. Sharafaldin, A. Habibi Lashkari, A.A. Ghorbani, "A Detailed Analysis of the CICIDS2017 Data Set," (2019). https://doi.org/10.1007/978-3-030-25109-3_9
13. A.H. Lashkari, G.D. Gil, M. S. I. Mamun, and A. A. Ghorbani, "Characterization of tor traffic using time based features," (2017). https://doi.org/10.5220/0006105602530262
14. Kurniabudi, D. Stiawan, Darmawijoyo, M.Y. Bin Bin Idris, A.M. Bamhdi, R. Budiarto, "CICIDS-2017 dataset feature analysis with information gain for anomaly detection," IEEE Access, (2020). https://doi.org/10.1109/ACCESS.2020.3009843
15. I. Sharafaldin, A. H. Lashkari, and A. A. Ghorbani, "Toward generating a new intrusion detection dataset and intrusion traffic characterization," 2018, doi: https://doi.org/10.5220/000663 9801080116.
16. S. Gamage, J. Samarabandu, Deep learning methods in network intrusion detection: a survey and an objective comparison. J. Netw. Comput. Appl. (2020). https://doi.org/10.1016/j.jnca. 2020.102767
17. H. K. K. Hyunjae Kang, D.H. Ahn, G.M. Lee, J.D. Yoo, K.H. Park, "IoT network intrusion dataset," IEEE Dataport, (2019). https://dx.doi.org/https://doi.org/10.21227/q70p-q449
18. Y. Meidan et al., "N-BaIoT-Network-based detection of IoT botnet attacks using deep autoencoders," IEEE Pervasive Comput (2018). https://doi.org/10.1109/MPRV.2018.03367731
19. A. Parmisano, S. Garcia, M.J. Erquiaga, "Stratosphere laboratory. a labeled dataset with malicious and benign IoT network traffic," Zenodo (2020)
20. T.M. Booij, I. Chiscop, E. Meeuwissen, N. Moustafa, F.T.H.den Hartog, "ToN_IoT: the role of heterogeneity and the need for standardization of features and attack types in IoT network intrusion datasets," IEEE Internet Things J (2021). https://doi.org/10.1109/JIOT.2021.3085194
21. A. Alsaedi, N. Moustafa, Z. Tari, A. Mahmood, A. Anwar, "TON_IoT telemetry dataset: a new generation dataset of IoT and IIoT for data-driven intrusion detection systems," IEEE Access (2020). https://doi.org/10.1109/access.2020.3022862
22. R. Damasevicius et al., "Litnet-2020: an annotated real-world network flow dataset for network intrusion detection," Electron (2020). https://doi.org/10.3390/electronics9050800
23. M.S. Elsayed, N.A. Le-Khac, A.D. Jurcut, "InSDN: A novel SDN intrusion dataset," IEEE Access (2020). https://doi.org/10.1109/ACCESS.2020.3022633
24. A. Kaan Sarica, P. Angin, "A Novel SDN Dataset for Intrusion Detection in IoT Networks," (2020). https://doi.org/10.23919/CNSM50824.2020.9269042

Chapter 6
Unsupervised Deep Learning for Secure Internet of Things

This chapter elaborates on the potential of unsupervised deep learning solutions for assuring the security of IoT-based systems to give the reader an insightful discussion of how these solutions could satisfy the IoT security requirements necessary to realize a reliable and trustworthy IoT environment. Unsupervised Learning is another common learning paradigm for deep learning models, which alleviates the need for supervision during the model training by allowing the model to be using its own to uncover hidden patterns and learn important data representations from unlabeled data [1]. The main goal of unsupervised learning is to capture high-order interrelationships from the noticed or discernible data for synthetic or analytical reasons under the absence of information about target classes. The potentials of deep unsupervised learning to discover relationships and variations in information render it a perfect solution for exploratory data analysis, cross-selling policies, client grouping, and similar complex tasks [2]. This chapter taxonomize unsupervised deep learning into three main categories, each of them is through the following sections.

- Generative adversarial Networks (GANs).
- autoencoder models
- energy-based models (EBMs)
- Summary and Learnt Lessons

6.1 Generative Adversarial Networks

The Generative Adversarial Networks (GANs) is broadly known unsupervised deep learning algorithm that falls under the umbrella of generative deep networks. Nevertheless, different from the autoencoders (AEs), generative networks are talented to generate new and profound results given random encodings [3] (see Fig. 6.1).

Electronic Supplementary Material The online version of this chapter (https://doi.org/10.1007/978-3-030-89025-4_6) contains supplementary material, which is available to authorized users.

M. Abdel-Basset et al., *Deep Learning Techniques for IoT Security and Privacy*, Studies in Computational Intelligence 997, https://doi.org/10.1007/978-3-030-89025-4_6

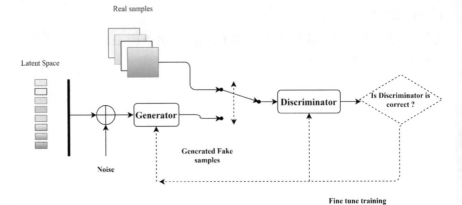

Fig. 6.1 Architecture of the generative adversarial network

GANs can learn how to model the input distribution by training two participating (and collaborating) deep architecture known as **generator** and **discriminator**. The generator takes the responsibility of discovering the way to engender fake input samples that could deceive the discriminator. In The Meantime, the discriminator is trained to differentiate real data from fake ones. With the advance of the training, the discriminator will no longer be capable to distinguish the real input from the synthetically generated input. In this way, the discriminator could be discarded, while the generator could be employed later to generate a new and different set of credible samples.

The fundamental notion of GANs is simple. Nonetheless, stabilizing the training of both generators and discriminators is deemed as a challenging facet to be considered. Hence, the generator and discriminator should have a nutritious contest to effectively learn at the same time. As expected, the loss is calculated using the discriminator's output to rapidly update its parameters. Hence, the rapid converges of discriminator make the generator no longer gets enough parameters' updates and unable to converge. In addition to being challenging to train, GANs could also undergo partial or complete modal breakdown, representing a condition in which the generator is delivering almost analogous results for various latent encodings [4].

In GAN, the generator receives noisy input and produces a synthetic sample as output. In The Meantime, the input of the discriminator can be either a real or a synthetic sample. Real samples were obtained from the genuine sampled data, whilst the fake samples are obtained from the output of the generator. All the legitimate samples are annotated as 1.0 while all the synthesized samples are annotated as 0.0. Because the annotation operation is automated throughout the training, GANs are yet deemed as an unsupervised deep learning approach. The discriminator takes the responsibility of learning how to differentiate fake samples from real ones. Throughout this portion of GAN training, merely the parameters of the discriminator are updated. Similar to a conventional binary classifier, the discriminator seeks to estimate on a scale of 0.0–1.0 in certainty values on in what way the generated samples are close to the real samples [5].

At routine periods, the generator makes as if its output is a legitimate sample and necessitate the GAN to be annotated as 1.0. When the false samples are then passed to the discriminator, instinctively it would be categorized as false with a probability score near 0.0. The optimizer calculates the updates of parameters of the generator network according to the offered label (which is, 1.0). Likewise, consider its own estimate while training using new samples. Specifically, the discriminator has some uncertainty regarding its output, and hence, GANs consider that during the training. To this end, GANs enable backpropagation of the gradients from the final layer of the discriminator down to the generator's opening layer. Nevertheless, in the majority of methods, throughout this training stage, the parameters of the discriminator are momentarily halted. Then, the GAN updates the parameters of the generator using the gradients information aiming to enhance its capability to generate fake samples. Generally, the entire procedure is similar to two models battling or playing against each other while remaining collaborating together. With the convergence of GAN, the final outcome is a generator able to synthesize samples similar to the real ones. The discriminator believes these synthesized samples are genuine or have an estimated label close to 1.0, which implies that the discriminator could then be abandoned as the generator is valuable in generating profound outputs from indiscriminate noise.

The training of discriminator seeks to optimize some loss function such as cross-entropy, in which the loss is computed as a negative sum of the expectancy of appropriately recognizing real data, and then subtracting the expectancy of properly detecting synthetic data from 1. Generally, the discriminator is fed with a couple of mini-batches during training, namely real samples labeled as 1.0, and synthesized samples with label 0.0. To optimize the loss function, the parameters of the discriminator are updated during the backpropagation by perfectly categorizing the real data and synthesized data. The generator receives a noise vector to synthesize the fake samples that contribute to optimizing the loss function. Hence, the GAN trains the generator by treating the total of the losses of the discriminator and generator like a zero-sum game, such that the loss of the generator is basically the negative of the loss of the discriminator.

In deep learning, the generator and discriminator are typically designed using an appropriate deep network. For example, in the case of image data, they both are designed with CNN. In the case of one-dimensional sequences, they are mostly designed using sequence-based models (recurrent networks, transformers, etc.). For graph data, the generator and discriminator are designed using graph networks [6].

6.2 Autoencoders

Autoencoder (AE) is defined as a popular unsupervised deep learning model where the output and input have identical dimensionality, which means that the number of output units is the same as the number of input units.

In the simplest style, the autoencoder can learn the inherent data representations by striving to generate output as a copy of its input. Nevertheless, utilizing an AE is not as straightforward as reproducing the input to output. Otherwise, the deep network will become unable to discover the hidden patterns in the input distribution. Typically, an AE model encodes the received input data into a low-dimensional or compressed representation that is ordinarily shaped as a vector. This will approximate the hidden structure that is generally known as the latent representation, latent space, or latent vector. This operation represents the encoding network. Then, the latent representations are passed to the decoder network to be decoded to the original input (see Fig. 6.2). In Particular, the AE imposes a bottleneck in the model which obliges a compacted representation of the initial input. When the features of input are autonomous, this compactness and following reconstruction will be a complicated operation. Nevertheless, when the data is structured in some way (i.e., contain correlated features), then, the input features could be easily learned and subsequently leveraged at what time obliging the input across the bottleneck.

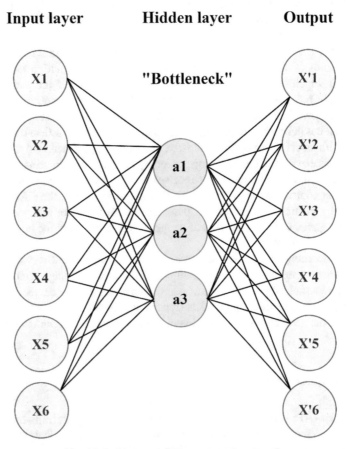

Fig. 6.2 Architecture of the auto encoder network

Like the earlier chapters, the input data can be any form of IoT data such as sensor data, time series, signals, text audio, video, image, etc. The main objective of AE is to discover and learn representation to achieve valuable input transformations. It is likely that the reproduced output by the decoder might only estimate the input. The difference between the reproduced output and input is estimated using a loss function.

However, why would we utilize autoencoders in IoT? For simplicity, the AE model has useful applications either in its original structure or as a portion of a more multifaceted deep learning model. Hence, it is considered a major tool for realizing advanced deep learning solutions since it could effectively handle and execute structural functions on the input data. Popular use cases incorporate denoising, tracking, colorization, feature-level mathematics, detection, and segmentation, etc.

To be more specific, the design of the AE model entails two operators:

- Encoder: it is responsible for transforming the input x, into a latent representation $z = f(x)$, that has a low dimension, hence, the encoder is obliged to just learn the most essential features from input. a typical example of that is handwritten digits recognition, in which the essential features to learn from input images might involve writing style, the roundness of stroke, angles, thickness, etc.
- Decoder: it is responsible for recovering the input using the latent representations $g(z) = \bar{x}$, Though the latent representation possesses a low dimension, it has an appropriate size to permit the decoder to reproduce the approximate version of original data. In other words, the objective of the decoder is to make $\bar{x}$ more similar to x.

Commonly, the encoder and decoder are implemented as non-linear function. Hence, they can be implemented using neural networks, i.e., RNN, MLP, CNN, etc. The small dimension of latent vector z represents the number of prominent features it captures. This dimension is often considered far smaller than the dimension of the input to construct the latent representation to capture merely the most prominent estates of the input distribution [7–9]. In contrast, when the dimension of latent space is larger than the input dimension, the AE model show tendency to memorize the input. An appropriate loss function, $L(x, \bar{x})$ should be employed to estimate the dissimilarity between input x, and reconstructed output $\bar{x}$. example of reconstruction loss function includes Mean Squared Error (MSE), binary cross entropy, structural similarity index (SSIM), etc. after setting an appropriate reconstruction loss function, the AE can be trained to optimize this function via backpropagation. Like other deep learning models, the only constraint for the loss function is that it should be differentiable. The default design of AE is implemented with MLP, however, the variations in the design of Encoder and Decoder parts bring us with different variants of AE discussed in the following subsections.

6.2.1   Sparse Auto Encoder (SAE)

Sparse autoencoders (SAE) present an alternate technique for establishing a latent representation without necessitating a decrease in the count of nodes in the hidden layers. Instead, SAE creates a loss function that punishes the activations of neurons in the network layers. For every provided observation, the network seeks to learn an encoding and decoding based on the activation of a minor number of neurons. This can be considered as a distinct method for regularizing the activations rather than the weights of the network. A generic SAE is envisioned below where the complexity of a node relates with the concentration of activation. It's essential to notice that the selection of the activated nodes of a trained model based on received input i.e., data-dependent, various inputs would result in activations of diverse nodes across the network (see Fig. 6.3).

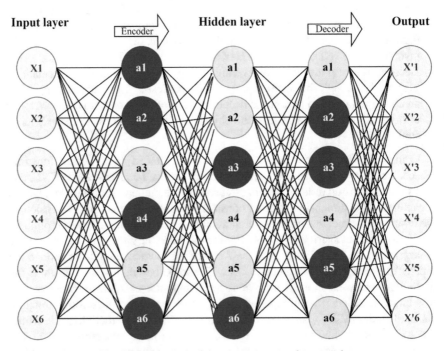

Fig. 6.3 Architecture of the sparse auto encoder network

Although an under-complete AE uses the whole network for each input sample, an SAE enables selective activation of parts of the network based on the input samples. Consequently, the capability of the network is constrained to memorize the input data without restricting the network's ability for feature extraction which enable considering the latent representation and network regularization *separately*, in such a way that the latent representation is determined according to the context of the data while using sparsity restraint to impose regularization. Two major methods exist for imposing such sparsity constraint namely L1-Regularization and KL-Divergence; both entail quantifying the activations of hidden layer for each training batch and inserting certain terms to the loss function to punish unnecessary activations.

6.2.2 Denoising Auto Encoder (DAE)

As previously discussed, the notion of training a network in which the input and outputs have the same dimension is achieved by an AE that seeks to reproduce the input as tightly as potential while passing across some kind of representational bottleneck. Bear In Mind that AE should be delicate as much as necessary to reconstruct the original samples but not excessively sensitive to the training data so that the model can learn a generalizable encoding and decoding. Another attitude in the direction of building a generalizable AE is to marginally distort the input samples but still retain the uncontaminated data as the required output. With this methodology, AE can't basically build a mapping that remembers the training data for the reason that the model's input and intended output become not identical. Instead, the AE model learns the mapping of input samples into a lower-dimensional representation; hence, when this representation correctly defines the real data, the AE essentially "canceled out" the injected noise (see Fig. 6.4).

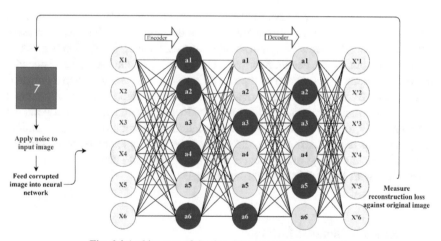

Fig. 6.4 Architecture of the denoising autoencoder network

6.2.3 *Variational Auto Encoder (VAE)*

A variational autoencoder (VAE) offers a *probabilistic* method for explaining the input samples in latent representation. Hence, instead of developing an encoder that yields a single value to define each feature of latent representation, the encoder is reformulated to explain a probability distribution for each latent feature (see Fig. 6.5).

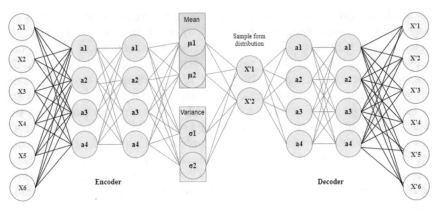

Fig. 6.5 Architecture of the variational auto encoder network

With this method, for each provided input, the corresponding *latent feature is represented* as a probability distribution. While decoding from the latent representation, the network selects a random sample from each latent distribution to produce a vector as input for the decoder network. Hint: For VAEs, the encoder and decoder network are sometimes known as the recognition network and generative network, respectively. By constructing the encoder network to produce an array of potential values from which a random sample is selected to be fed to the decoder network, the VAE basically imposing a constant, smooth latent representation. For any random sample of the latent distributions, the decoder network is expected to be capable of precisely recreate the input. Hence, adjoining values in latent space have to be compatible with almost identical reconstructions.

6.3 Energy-Based Models

The EBMs try to encode the interdependencies between the input samples and the hidden neurons by learning the mutual probability. This kind of model is consisting of multiple neurons with bidirectional connection in-between, which may be divided into neurons denoting the input variables and neurons denoting the hidden variables and may follow either Gaussian distribution or Bernoulli distribution. Each setting of the neurons and energy estate could be delineated and encoded in the weights of the neurons. The model learning process is constrained by an effort to reduce the overall energy of the deep network, a condition that is achieved once its energy is contrarywise relative to its joint probability, in accordance with the distribution of the training samples. In the subsequent subsection, the discussion underlines some common EBMs, including Boltzmann machines (BM), restricted Boltzmann machines (RBMs), deep Boltzmann machines (DBMs), deep belief networks (DBNs), and other pertinent extensions.

6.3.1 Boltzmann Machine (BM)

Boltzmann machines were first introduced by Geoffrey Hinton and Terry Sejnowski in 1985. The BM can be defined as a common unsupervised network consisting of regularly connected neurons that are stochastically activated via a straightforward learning algorithm that tolerates them to uncover noteworthy features in the input data. Boltzmann machines—of the unlimited kind—comprise a neural network with a single input layer as well as one or more hidden layers. The neurons of BM layers are responsible for making stochastic decisions regarding whether they should activate or deactivated depending on the context of data they receive during training and the loss function the BM attempting to optimize. In this way, the BM uncovers the fascinating features regarding the data, which facilitates modeling the multifaceted inherent interrelationships and patterns embedded in the input data. Nevertheless, these unrestricted BM (UBM) enable the neurons of any layer to connect to a neuron from the other layers of neurons in the same layer. Accordingly, with the existence of numerous hidden layers, the training of UBM becomes ineffective and computationally exhaustive. The learning of BMs can be accelerated by setting one layer of feature detectors per time (see Fig. 6.6).

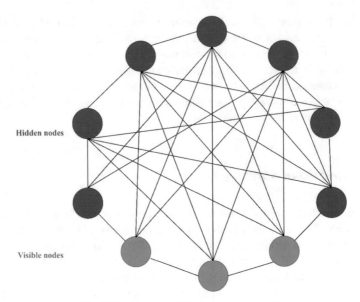

Fig. 6.6 Architecture of the boltzmann machine

6.3.2   Restricted Boltzmann Machine (RBM)

To address the limitation of UBM, a restricted Boltzmann machine is introduced in 1986 as a special case of BM, which belongs to a type of unsupervised network of evenly connected neuron-similar units that are cable to take a stochastic decision. RBM is defined as a two-layered generative model that has the capability to learn a probability distribution of input samples to detect automatically inherent representations in the data via input reconstruction. The connection between the neurons in the same layer is forbidden to facilitate the learning process. This restriction makes the hidden units are tentatively autonomous given an input vector, so unbiased representation can be extracted from its input sample single step. Again, each neuron act as a venue of computation that handles the input and starts by getting stochastic decisions concerning whether to communicate that input or not (see Fig. 6.7)

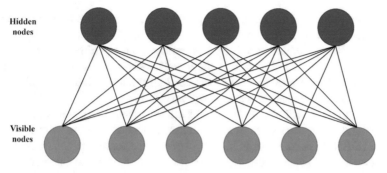

Fig. 6.7 Architecture of the restricted Boltzmann machine

6.3.3 Deep Boltzmann Machine (DBM)

Deep Restricted Boltzmann machine (DBM) can be viewed as a deep-designed version of RMBs in which hidden neurons are assembled under a hierarchical layer rather than a single layer. Similar to the connectivity restriction of RMB, full connectivity is only present between successive layers, and no connections within layers are permitted. DBM seeks to perform hierarchical learning of the features from the input data, whereas the features captured in the particular layer are directed as an input to the following layer in form of hidden variables. Unlike RBM, DBM establishes extra connections between latent features to create a multi-layer structure in the hidden portion of BM. The latent representations are detected via the utilization of DBM consisting of unidirectional connections. The greedy layer is employed to carry out a valuable learning and parametric interpretation process. The best possible parameters are discovered utilizing the unsupervised representation. This deep neural architecture enables learning various degrees of representations from the input observations according to multi-layer structures.

6.3.4 Deep Belief Network

Deep learning has been witnessing increased popularity in the artificial intelligence community since the revolutionary study introduced in 2006 by Hinton et al. [10], which presents a new methodology for effective training of deep networks. The key challenges causing ineffective training stem from the arbitrary parameter initializations and conventional gradient optimization, which had formerly been known to cause training collapse in the majority of deep networks. Hinton et al. instituted the right technique for effective representation learning exploiting a complementarity in advance of defeating the abovementioned challenges.

Motivated by the perception of deep architecture as a chronological model, a DBN was presented as a probabilistic model formed by chronologically merging several simple blocks that are obliged to extract a variety of representations. The learning procedure of DBN follows the greedy layer-wise training, in which one layer is trained per timestep. From a structure viewpoint, a DBN is similar to the multilayer perceptron, nevertheless, they vary in their training methodology (see Fig. 6.8). Similar to RBMs, DBNs could learn the inherent representations of input samples and reconstruct them in a probabilistic manner. Also, as with RBMs, the layers in DBNs have connections that only connect the neurons from different layers. DBN is a hybrid generative network that consists of manifold-directed sigmoid-activated layers with an exception for the first and last layer, which establishes an RBM model; the multiparty distribution of the features in the layers of a DBN may be encoded into the provisional probabilities of features within two consecutive layers. The training of a DBN is accomplished via a recursive method that begins with RBM's training. Then, the input-to-hidden weights are hidden layers are preserved, and their replica is employed to assign initial values to the weight matrix of a new layer. The hidden features of the initial layer then turn out to be the input features of the next layer, and the training is carried out using the initialized weight matrix.

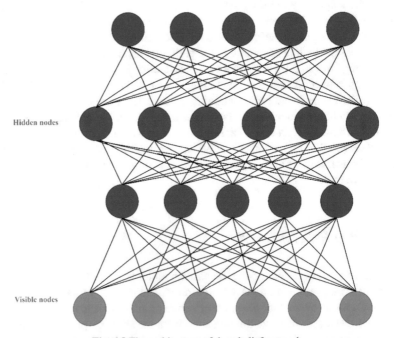

Fig. 6.8 The architecture of deep belief network

6.4 Summary and Learnt Lessons

This chapter provides an in-depth discussion about the role of unsupervised deep learning that could provide efficient security solutions for IoT environments, especially in unlabeled data scenarios. In the nutshell, this chapter delivers useful insights regarding the different unsupervised deep learning as follow:

- This chapter discusses different categories of an unsupervised deep learning solution with the main aim to understand the different methodologies to learn and extract the valuable representations from data samples. The argued models and the chapter organization are decided to comply with the requirements of modern IoT applications, where the data size is bursting accompanied by high-dimensional samples.
- The exhaustive amount of daily generated unlabeled data reveals that the capability of the unsupervised deep learning solution to detect attacks/intrusions highly relies on the nature of the attack and on the number of classes taken into account during the training phase.
- Unsupervised deep learning is taxonomized into three primary categories i.e., energy-based, generative, and autoencoders. Each of these categories is divided into more subcategories. This taxonomy aims to help beginners' researchers to realize the characteristics of unsupervised approaches and their promising role in securing IoT-based systems.
- The application of generative methods shows great success in vision-related data. however, investigating them for IoT data such as sensory data, wearable, traffic flows is still in the early phases. unlike supervised security techniques, unsupervised solutions can identify ransomware with interactive transformations and ransomware without labeling burdens.

References

1. G. Wilson, D.J. Cook, "A survey of unsupervised deep domain adaptation." ACM Trans. Intell. Syst. Technol. (2020). https://doi.org/10.1145/3400066
2. G.J. Qi, J. Luo, "Small data challenges in big data era: a survey of recent progress on unsupervised and semi-supervised methods." IEEE Trans. Pattern Anal. Mach. Intell. (2020). https://doi.org/10.1109/tpami.2020.3031898
3. Z. Wang, Q. She, T. Ward, "Generative adversarial networks in computer vision: a survey and taxonomy." ACM Comput. Surv. (2021). https://doi.org/10.1145/3439723
4. D. Saxena, J. Cao, "Generative adversarial networks (GANs)." ACM Comput. Surv. (2021). https://doi.org/10.1145/3446374
5. Y. Deldjoo, T. Di Noia, F. A. Merra, "A survey on adversarial recommender systems: from attack/defense strategies to generative adversarial networks." ACM Comput. Surv. (2021). https://doi.org/10.1145/3439729
6. D. Valsesia, G. Fracastoro, E. Magli, "Learning localized representations of point clouds with graph-convolutional generative adversarial networks," IEEE Trans. Multimed. (2021). https://doi.org/10.1109/TMM.2020.2976627

7. L. Zheng, B.J. Hu, J. Qiu, M. Cui, "A deep-learning-based self-calibration time-reversal finger-printing localization approach on Wi-Fi platform." IEEE Internet Things J. (2020). https://doi.org/10.1109/JIOT.2020.2981723
8. F. Gu, K. Khoshelham, C. Yu, J. Shang, "Accurate step length estimation for pedestrian dead reckoning localization using stacked autoencoders." IEEE Trans. Instrum. Meas. (2019). https://doi.org/10.1109/TIM.2018.2871808
9. M. Kim, D. Han, J.K. Rhee, "Multiview variational deep learning with application to practical indoor localization." IEEE Internet Things J. (2021). https://doi.org/10.1109/JIOT.2021.3063512
10. G. E. Hinton, S. Osindero, Y. W. Teh, "A fast learning algorithm for deep belief nets." Neural Comput. (2006). https://doi.org/10.1162/neco.2006.18.7.1527

Chapter 7
Semi-supervised Deep Learning for Secure Internet of Things

Previous chapters demonstrate the great success achieved by principally in supervised settings, by leveraging a larger volume of precisely annotated dataset. Nevertheless, annotated data instances are frequently complicated, costly, or laborious to acquire. The annotation procedure typically necessitates many experts' efforts, which is the foremost drawback for training a brilliant supervised deep model. when a small number of annotated observations are obtainable, it could be interesting to develop a thriving learning solution. On the opposite side, the unlabeled data is generally plentiful and could be simply or cheaply attained. Therefore, it is advantageous to use a huge number of unannotated data for enhancing the learning performance given a little amount of annotated data. motivated by that, semi-supervised learning is regarded as an interesting research field concerned with developing efficient models trained with both labeled and unlabeled data. it also seeks to improve the learning performance by leveraging extra unlabeled examples contrasted to supervised approaches based merely on the annotated samples. semi-supervised solutions could easily obtain by extending supervised or unsupervised learning algorithms. Therefore, this chapter emphasis discussing the potential of a state-of-the-art deep learning solution that can be exploited for securing the Internet of Things (IoT) by leveraging partially labeled data.

This chapter covers the following topics.

- Background and Foundations
- Consistency Regularization for Semi-supervised Learning
- Semi-supervised Generative Approaches
- Semi-supervised Autoencoder Approaches
- Semi-supervised Graph-Based Approaches
- Hybrid Approaches
- Summary and Learnt lessons

Electronic Supplementary Material The online version of this chapter (https://doi.org/10.1007/978-3-030-89025-4_7) contains supplementary material, which is available to authorized users.

181
M. Abdel-Basset et al., *Deep Learning Techniques for IoT Security and Privacy*,
Studies in Computational Intelligence 997,
https://doi.org/10.1007/978-3-030-89025-4_7

7.1 Background and Foundations

This section aims to introduce the formulate the fundamentals of semi-supervised deep learning. Given $X = \{X^L, X^U\}$ representing the total dataset containing a small number of annotated samples $X^L = \{(x^i, y^i)\}_{i=1}^{L}$ as well as a large number of unannotated samples $X^U = \{(x^i)\}_{i=1}^{U}$ i.e. $L < U$. Given that the dataset contains K categories, where the earliest L instances of the X are annotated with $\{(y^i)\}_{i=1}^{L} \in \{y^1, y^2, \cdots, y^K\}$, then the semi-supervised training of deep networks can be designated as optimizing the following loss:

$$\min_{\theta} \underbrace{\sum_{(x,y) \in X^L} L_s(x, y, \theta)}_{\text{supervised loss}} + \alpha \underbrace{\sum_{(x) \in X^U} L_u(x, \theta)}_{\text{unsupervised loss}} + \beta \underbrace{\sum_{(x) \in X^U} \mathcal{R}(x, \theta)}_{\text{regulariztion loss}} \qquad (7.1)$$

where L_s represent the supervised loss per sample, i.e., Dice loss for segmentation, cross-entropy for classification, L_u and $\mathcal{R}$ represents the unsupervised and regularization loss per sample. Regularization loss is also known as consistency loss or regularization term. Bearing in mind that terms of unsupervised loss are habitually not rigorously differentiated from regularization terms, because both of them are not handled by label information. Finally, θ represents the parameters of the deep network, α, and β represent the scalar weight for balancing different loss terms. A variety of options of the unsupervised and regularization loss bring us with a variety of semi-supervised algorithms.

7.1.1 Semi-supervised Learning Assumptions

Semi-supervised learning seeks to offer efficient learning performance leveraging both annotated and unannotated data samples aiming to overcome the limitations of the reliance of supervised learning on large amounts of annotated data. Nevertheless, an indispensable precondition is that the sample distribution has to follow some assumptions. When this is not satisfied, semi-supervised might not get better than supervised learning and even destroy the learning performance via deceptive inferences [1]. By the way, the assumptions of semi-supervised deep learning could be formulated as follow:

7.1.1.1 Semi-supervised Smoothness Assumption

When two data elements $x1$ and $x2$ are closed to each other in a high-density region, then their predictions $y1$ and $y2$ should also be close to each other [2]. This proposition indicates that when two examples belong to a particular cluster, their predictions would have a tendency to belong to the same class label. In contrast, when the samples are divided through a low-density area, and later the corresponding predictions have a tendency of being dissimilar.

7.1.1.2 Cluster Assumption

When two data elements $x1$ and $x2$ are belonging to one group, they have to fit in the consistent class. This supposition describes the fact that data elements of a particular class have a tendency to create a group, and as soon as the elements could be linked by small curves that did not move across any low-density zones, they are part of the identical class group [2]. Consequently, the decision boundary has to not traverse high-density zones but as an alternative stay in low-density zones [26]. Following this assumption, the intelligence driven solution is capable of using a significant number of unannotated samples to adapt the categorization frontier.

7.1.1.3 Low-Density Separation

In this scenario, the classification margin ought to belong to a low-density zone, instead of high-density zone. Clustering supposition could be deemed from an alternative standpoint by pretending that the category is divided by regions of low density [2]. While the classification margin in a high-density zone will split a group into a couple of separate classes and the resultant portions will contravene the cluster hypothesis.

7.1.1.4 Manifold Assumption

The data samples fall into a small-scale manifold. When couple of data elements $x1$ and $x2$ are situated in a local region in the small dimensionality multifarious, they have comparable class labels. This hypothesis indicates the regional softness of the decision boundary. It is popular that the curse of dimensionality is one of the most crucial challenges in artificial intelligence. It is difficult to approximate the genuine data distribution once capacity expands exponentially with in high dimension settings. When the samples fall on a low-dimensional multifarious, the intelligence driven solution could escape the scourge of dimensionality and function well in its low-dimension domain. The leading distinction that distinguishes this assumption from the cluster assumption is that the latter one primarily concentrates on global distribution, while the former one principally contemplates the vicinity of the data [3].

7.1.2   Related Theories

7.1.2.1   Transfer Learning

Transfer learning is defined as a learning paradigm that seeks to transfer the experience from single or multiple domains (known as source) to a particular domain (called target) with the aim of improving performance on the target task. Different from semi-supervised learning, which performs perfectly under the premise that the input data are independent and identically distributed (IID), transfer learning enables the tasks, spaces, and deliveries exploited for learning and inferencing to be changed however associated [4].

7.1.2.2   Weakly Supervised Learning

Weakly supervised learning is a strategy for easing the data reliance that necessitates an manual annotation mask to be provided for a substantial quantity of training samples under intense supervision. In this regard, three categories of weakly supervised data are present in the literature namely incomplete data, inexact data, and inaccurate data. Incomplete categories imply that only a part of training dataset was annotated. Under these settings, domain adaption and semi-supervised approaches are considered as leading solutions. In the inexact data category, the data samples have a coarse-grained annotation, hence multi-instance learning is viewed as a leading approach for this case. For inaccurate data category, the samples are given labels that are not constantly considered as a mask, like the condition of noisy-efficient training [5].

7.1.2.3   Meta-Learning

Meta-learning is an improved concept commonly referred to as "learning to learn", which seeks to facilitate learning additional abilities or acclimate to different responsibilities promptly based on the earlier knowledge acquired from a few training instances. It is popular that artificial intelligence algorithm often necessitates a huge amount of training samples. The meta-learning approaches are anticipated to acclimate and perform generalization to different settings which is usually confronted throughout the learning practice. The adjustment procedure is basically a mini learning period that takes place in the course of the testing but has restricted coverage to newfound task configurations. Finally, the tailored scheme could be trained on a variety of learning operations and boosted on the distribution of events, involving possibly unwitnessed tasks [6].

7.1.3 Taxonomization

The current literature contains a broad range of semi-supervised approaches such as generative networks, self-ensembling, graph-based approaches, self-training [22]. This makes the study of semi-supervised deep learning more complex and difficult to deliver a full view for the readers. Motivated by that, this chapter taxonomizes the state-of-the-art semi-supervised into five main subcategories according to the loss functions and as well as model architecture. This includes Consistency Regularization Approaches, Semi-supervised Generative Approaches, Semi-supervised Autoencoders Approaches, supervised graph-based Approaches, and hybrid approaches. Each of these categories can be further divided into subcategories and the discussion of each category is argued in the following sections. Figure 7.1 shows the taxonomy of the state-of-the-art semi-supervised deep learning approaches.

7.2 Consistency Regularization Approaches

This section discusses and presents the consistency regularization approaches as a common category of semi-supervised deep learning. The concept of consistency regularization usually applies some objective function for determining the preceding restrictions presumed by scholars. Consistency regularization often depends on the smoothness hypothesis and/or the manifold hypothesis and defines a kind of approach in which the sensible transformations of the input elements never affect the performance of deep network [7]. Therefore, consistency regularization could be considered as a method to discover a suave manifold upon which the training data rests by

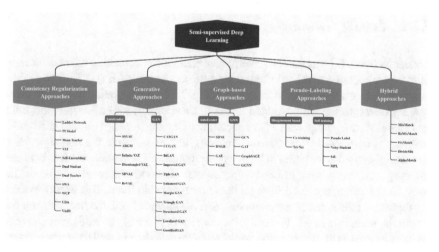

Fig. 7.1 Taxonomy of semi-supervised deep learning approaches

exploiting the unannotated data samples [8]. The highly popular architectural design of consistency regularization approaches follows the Teacher-Student network architecture. Where the student network learns in a normal way, whilst the teacher network takes the responsibility of generating targets. Given That the model itself is responsible for target generation, the targets might be erroneous and later utilized by the student network for learning purposes. Essentially, the consistency regularization approaches struggle with verification bias, a problem that could be alleviated by enhancing the quality of generated targets. Officially, it could be presumed thar the training dataset X contain two sets of elements namely annotated data X^L and unannotated data X^U. Given θ and θ' respectively representing the parameter of rudimentary student network and the target, then, the consistency constraint can be formulated as:

$$\mathbb{E}_{x \in X} \mathcal{R}(\mathbf{f}(\theta, x), \mathcal{T}_x) \tag{7.2}$$

whereas $f(\theta, x)$ represent the output estimated by from student network $f(\theta)$ based on input x.

$\mathcal{T}_x$ denote the constancy target from the teacher network. $\mathcal{R}(;)$ is a function that calculating the dissimilarity between two vectors and is mostly designated with KL-divergence or Mean Squared Error (MSE). A variety of consistency regularization approaches differ in the way they synthesize the targets. There are numerous methods to enhance the quality of the generated target $\mathcal{T}_x$. One method tries to decide on the perturbation instead of multiplicative or additive noise thoroughly. Another method is to carefully contemplate the teacher network rather than reproducing the student network.

7.2.1 Ladder Network

Ladder Network [9] was the earliest effective effort towards making use of a Teacher-Student architecture that is motivated by a deep denoising AE (DAE). The Ladder network 's structural design is illustrated in Fig. 7.2. In the encoding part, some noise was injected into the whole hidden layers in form of a distorted feedforward route that shares the mappings with a pure encoding of the feedforward route. The decoder pathway is composed of particular denoising functions, while the non-supervised denoising squared error per each layer was contemplated as consistency loss between the output of every layer in the two pathways. The ladder network can be distinguished from standard DAE through the underlying skip links. This aspect enables the features of the upper layer to emphasize more encapsulated invariant characteristics in the underlying task. Formally, the consistency loss of the ladder network can be calculated as MSE for the activation of the pure encoding as well as the reproduced activations.

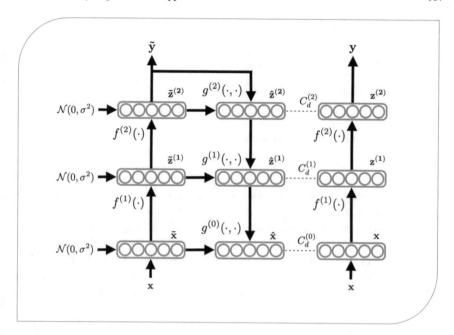

Fig. 7.2 The structure of the ladder network

7.2.2 Π-*Model*

Different from the perturbation presented in the Ladder Network, Π-Model [10] is introduced to perform an arbitrary transformation of input for both annotated as well as unannotated data. Certain methods with indeterministic conduct, like dropout, randomized augmentation, and indiscriminate max-pooling, enforce the input instances to pass across the network multiple times, resulting in a variety of predicted outputs. For every training epoch of Π-Model, the same unannotated data samples circulate ahead two times, while indiscriminate transformations are launched by means of dropout and data augmentations. This repeated forward propagation of input samples might cause various predictions, and the role Π-Model is to optimize consistency loss that make the network predictions more consistent under different input perturbations.

7.2.3 *Temporal Ensembling*

Like Π-Model, Temporal Ensembling (TE) [11] is introduced to perform a compromise prediction under various input augmentation circumstances and regularization. Figure 7.3 shows the architecture of the Temporal Ensembling framework. It can be considered as an extension to Π-Model that leverages the Exponential Moving

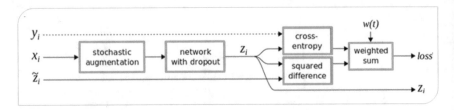

Fig. 7.3 The architecture of the temporal ensembling framework

Average (EMA) of predictions of earlier epochs. Which means, whilst Π-Model seeks to repeat the forward propagation of the input sample two times per every training iteration, TE decreases this computational burden by leveraging the EMA to aggregate the predictions through training epochs. In Particular, the ensemble productions Z^i is upgraded with the model predictions z^i following every training iteration, $Z^i \leftarrow \alpha Z^i + (1 - \alpha)z^i$, whereas α denote the momentum constant. Throughout the training procedure, the Z^i could be deemed to encompass an average ensemble of the outputs of the network because of stochastic augmentations and dropout.

7.2.4 Mean Teacher

As stated, TE retains an EMA of output estimates for each training instance and punishes estimates which are incompatible with that target. Nevertheless, since the targets vary only one time per epoch, the TE comes to be cumbersome when learning huge datasets, often present in IoT environments. Mean Teacher (MT) [12] was introduced to address this issue by calculating the averages network weights utilizing EMA throughout training iterations rather than label predictions. In this way, the MT tends to produce a further precise network rather than directly exploiting the predicted outputs, while training with a smaller number of labels than TE. Figure 7.4 shows the architecture of the MT framework which is composed of two networks termed

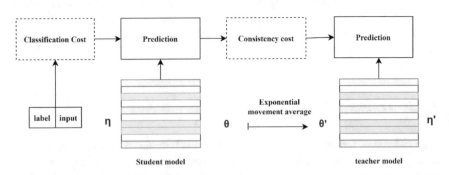

Fig. 7.4 The architecture of mean teacher framework

as Student and Teacher network. The student network is a standard network like the Π-Model, and the teacher network is the same structural design of the student network accompanied with EMA of the parameters of the student network. Next, MT is employed to validate regularization restraints between the estimated output of the teacher network and student network.

7.2.5 Dual Student

As an extension to the MT model, the Dual Student [13] approach was introduced by substituting the teacher network with an alternative student network. Figure 7.5 shows the architecture of the dual student framework in form of a Student–Student network. The dual student networks begin with various initial conditions and are updated throughout specific pathways in the course of training. This design brings us with a narrative notion termed as, "stable sample", accompanied by a equilibrium restrictions for evading the performing jam emitted by a paired EMA approach. Therefore, their weights might not be strictly paired, and every one of them can have self-knowledge according to its own learning. Properly, dual student approach seek to test out whether x represent a steady example for student.

7.3 Semi-supervised Generative Approaches

As debated in the previous chapter, the generative adversarial networks (GANs) could learn the inherent features of the input data by modeling the distribution of real unlabeled training data while generating new instances belonging to the same distribution. Motivated by this fact, the GANs can be exploited to empower the

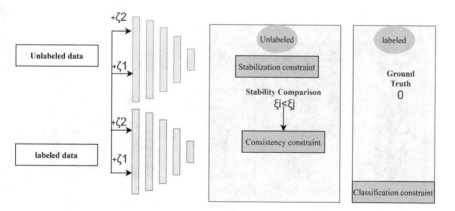

Fig. 7.5 The architecture of dual student framework

capabilities of semi-supervised learning. In this regard, four major scenarios could be noted for the usage of GAN for semi-supervised training (1) the reusage of features from the discriminator; (2) exploiting GAN-generated examples for classifier regularization; (3) training an inference network, and (d) exploiting the GAN-synthesized samples as supplementary training data. A straightforward semi-supervised approach can be present by combining the supervised loss and unsupervised loss throughout training, which has been demonstrated useful for many tasks i.e., classification, detection, etc. This observation shows that a straightforward and effective semi-supervised approach could be supplied by merging a supervised cost function and an unsupervised GAN function [14].

7.3.1 Categorical Generative Adversarial Network (CatGAN)

The CatGAN [15] adjusts the loss function of the standard GAN's to consider the joint information between commented data samples and their forecasted unconditional distributions of class. Specifically, the learning trouble is distinct from the standard GAN (previous chapter). The architecture of the CGAN is displayed in Fig. 7.6. This algorithm seeks to train the discriminator to differentiates the input instances into K classes by categorizing y for each x, rather than training a binary discrimination objective function. Additionally, the discriminator of CatGAN has its supervised loss design as a cross-entropy between the estimated restrictive distribution $P(y|x; D)$. The genuine label distribution of instances involves of three components, first, entropy $H[P(y|x; D)]$. employed to get specific class appointment for instances, second, $H[P(y|G(z); D)]$ employed for ambiguous estimates from synthesized instance; third, the secondary class entropy $H[P(y|D)]$ to regulate the exploitation of all categories. The CatGAN exploits the latent features acquired through the discriminator for the decisive training operation. For the supervised part

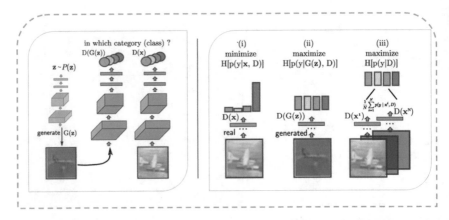

Fig. 7.6 The architecture of the CGAN

of data, the cross-entropy was employed supervised loss that considers the difference estimated outputs and the actual labels of training samples.

7.3.2 Context-Conditional Generative Adversarial Networks (CCGAN)

The CCGAN [16] was introduced to apply an adversarial loss to leverage unanno-tated image samples according to image in-painting. The primary focal point of this approach is the context information offered by means of the adjoining portions of the image. The approach start the training of the generator network to synthesize pixels within a erroneous gap. At the same time, the discriminator is presented to differentiate between the actual unannotated samples and those in-painted instances. Specifically, $m \odot x$ is passed as an input to a generator, where m represent a binary mask to abandon a specific section of an input image and $\odot$ represents Hadamard product. Hence, the in-painted image $x^I = (1 - m) \odot x_G + m \odot x$, where the output of generator $x_G = G(m \odot x, z)$. The in-painted samples supplied with the gener-ator lead the discriminator to capture the representations empowering the network to generalize to other associated classification tasks. The one before the last tier of elements of the discriminator is subsequently communicated with the learner having a cross-entropy cost function to optimize the model in combination with the discriminator loss.

7.3.3 GoodBadGAN

GoodBadGAN [17] assumes that the generator and discriminator might not be ideal when trained at the same time, i.e., the discriminator realizes an efficient semi-supervised learning performance, while the generator might engender visu-ally impractical examples. The approach provides hypothetical explanations of the reason behind utilizing poor examples from the generator to improve the learning performance. Commonly, the generated data instance might compel the periphery of the discriminator to be situated between the manifolds of data of various kinds, in such a way to enhances the generalizability of the discriminator network. As a result of the analysis, GoodBadGAN is shown to train depraved generators by obvi-ously injecting a punishment variable to produce corrupt examples, whereas some indicator function is employed to make certain that just high-level-density examples are disciplined while low-level examples are unimpacted. Furthermore, to ensure the robust true–false confidence in the optimum circumstances, a conditional entropy variable is included in the loss function of the discriminator.

7.4   Semi-supervised Autoencoder Approaches

Multiple justifications for considering latent-encoding representation algorithms valuable for efficient semi-supervised solutions. First, it provided a normal manner to include unannotated data for training. Second, the capability to extricate representations through the configuration of latent features. Third, they also enable the usage of variational networks. In this regard, variational autoencoders (VAEs) [18, 19] are elastic networks that bring together autoencoders and generative latent-space algorithms. The generative network model the inherent representations of the data distributions instead of the data observations and designate the mutual distribution to be $p(x, z) = p(z)p(x, z)$, whereas $p(z)$ denote the preceding distribution over latent representation z. Given That the correct posterior $p(x, z)$ is mostly troublesome, the training of generative network is assisted by an estimated posterior distribution $q(x, z)$. The structural design of VAEs follow a two-phase network, an encoder subnetwork to build a variational approximation $q(x, z)$ to the posterior $p(x, z)$, while the decoder subnetwork employed to regularize the probability $p(x, z)$.

7.4.1   Semi-supervised VAEs (SSVAEs)

The SSVAE approach [20] presents a semi-supervised solution depending on generative deep networks making use of two VAE-based generative networks to learn latent representations from input data. The discriminator network of latent feature space, termed M1 [20], could deliver more vigorous latent representations using the generative deep network. It is worth mentioned where $p_\theta(x, z)$ represent as an in-linear perturbation i.e., a deep model. Hidden feature space z could be nominated as a Bernoulli or Gaussian distribution. An estimated example of the posterior distribution on the latent features $q_\phi(x, z)$ is employed to act as a feature classifier targeting class y. Generative semi supervised network known as M2 [20] was presented to defines the data engendered with a latent feature y and a constant latent representation z, and is formulated as $p_\theta(x|z, y)p(z)p(y)$, which follow the multinomial distribution, wherever the actual category y are regarded as latent features of unannotated samples $p_\theta(x|z, y)$ denote an appropriate probability function. Missing labels can be forecasted using the implied posterior distribution $p_\phi(x|z, y)$. Packed semi-supervised generative network named $M1 + M2$, utilizes the generative network $M1$ for extracting the latent features $z1$, and utilizes the implanting from $z1$ rather than from input data x to train a semi-supervised generative network $M2$.

7.4.2 Infinite VAE

Infinite VAE [21] approach was introduced to combine VAE and nonparametric Bayesian approach in a single framework that could acclimate to outfit the input data by socializing parameters via a Dirichlet operation. It integrates Gibbs's sampling with variational inference to enable the network effectively to learn the patterns of the underlying input. in other words, the infinite VAE utilizes the blending coefficients to improve semi-supervised learning by leveraging the generative network (unsupervised) and a discriminator network (supervised), where the blending parameters were derived from a Dirichlet distribution, while the latent variables z^i in every VAE network were derived from a Gaussian distribution.

7.4.3 Disentangled Variational Autoencoder

Disentangled VAE approach [22] tried to model the disentangled representations utilizing partly stated graphical network and distinctive encoding facets of the dataset into distinct features. It investigates the graphical network for learning the global dependence on discerned as well as overlooked latent representations via deep networks, and a stochastic calculation graph [23] is employed to deduce and train the ensuing generative network. To this end, significance selection assessments are employed to augment the smaller limit of both the semi-supervised and supervised probabilities. Specifically, this approach takes into consideration the conditional probability implying that the model cannot calculate a straightforward Monte Carlo approximator via absolute distribution sampling. Hence, the variational shorter limit for supervised tenure increase toward less.

7.5 Semi-supervised Graph-Based Approaches

A Number Of efficient deep embedding approaches are developed to solve chief limitations of autoencoder-based approaches by constructing certain utilities that depend on the node's local neighborhood, however not inevitably the complete graph. The graph neural network (GNN), which is extensively employed in cutting-edge deep embedding methods, and could be considered as a typical baseline for the characterization of deep learning on the graphs. Similar to other deep node-level embedding techniques, a classifier was learnt to envisage the correct category for the annotated nodes [24]. Later, it could be employed to the unannotated nodes according to the decisive concealed status of the graph-centered network. Given that GNN is composed of two major functions: the aggregation function and the update function, the fundamental GNN along with some common extensions are discussed for semi-supervised settings a shown in the following subsections.

7.5.1 Baseline GNN

The essential facet of a baseline GNN [25] is that the advantage of neural message exchange is to swap and renovate the information between every pair of nodes utilizing deep networks. In particular, a hidden embedding per every neural message exchange throughout the GNN primitive epoch is upgraded based on the information from the locality in relation to every node. It is notable that the update and aggregate functions ought to normally be differentiable as with a standard neural network. A novel state is then engendered as soon as the locality message is merged with the previous concealed embedding state. Following several epochs, the final hidden embedding state reaches a convergence state so as to create the final state of each node as output. Principally, the information from the adjacent nodes is encapsulated in the beginning. Then, the locality information and the output of the preceding hidden node are combined via an indispensable linear mixture. After All, the combined information applies an activation function for introducing non-linearity. It is worth mentioning that the layers of GNN could be simply packed together, where the output of the final layer in GNN is considered as the definitive node embedding outcome to learn certain classifiers for the downstream tasks [26].

7.5.2 Graph Convolutional Network (GCN)

As early stated, the highly straightforward locality aggregation function only computes the total of the locality encoding conditions. The essential problem facing this approach is that nodes with a huge intensity seem to gain a superior value from more neighboring nodes in comparison with those having few numbers of neighboring nodes. One standard and simple approach to address this dilemma is to regulate the aggregation procedure based on the predominant node intensity. The extremely prevalent technique is to apply symmetric normalization. GCN completely utilizes the standardized locality grouping practice. Thus, the GCN model defines the update function without any aggregation strategy is defined. A huge scale of GCN variations is offered to improve the semi-supervised learning capabilities from different perspectives [27].

7.5.3 Graph Attention Network (GAT)

Beyond and above the typical approaches of performing aggregation, an additional popular approach for enhancing the aggregation layer of GNNs was achieved by applying particular attention procedures. The fundamental principle is to give out each neighboring node a score or importance ratio which is applied to estimate the impact of these neighboring nodes throughout the aggregation procedure. Hence, GAT was introduced to apply attention weights to characterize a weighted adjoining volume [28].

7.6 Pseudo-Labeling Approaches

7.6.1 Disagreement-Centered Methods

The concept of disagreement-based semi-supervised learning revolves around training manifold learning networks for a specific task and leverage the disagreement in the course of the training operation [29]. In this design scheme, two or more distinct deep networks are trained at the same time and assign labels for unannotated samples of each other. Disagreement-centered methods vary in whether the data maintains various interpretations i.e., Tri-Net [30] for particular-view data or Co-training [31] for multi-view data [32].

7.6.2 Self-Training Methods

In self-training approaches, the convinced predictions of network were leveraged to generate the pseudo annotations for unannotated data. this simply means that the model could include more training samples by utilizing present annotated data to forecast the labels of unannotated samples. Thanks to its straightforwardness and generalization, self-training is effectively applied in a variety of tasks including contour detection, Named Entity Recognition, object detection, and machine translation.

7.7 Hybrid Approaches

The Hybrid approaches integrated take the advantage of the before-mentioned approaches such as consistency regularization, pseudo-label, and generative models with the main aim to improve the performance of learning performance. in addition, a learning theory called Mixup [33] is presented and adopted by this category of deep semi-supervised approaches. the theory could be deemed as a straightforward, data-agnostic augmentation method, that is u-shaped grouping of coupled examples and the corresponding ground-truth. Strictly, Mixup creates simulated training instances,

$$\tilde{x} \leftarrow \lambda x^i + (1 - \lambda) x^j \tag{7.3}$$

$$\tilde{y} \leftarrow \lambda y^i + (1 - \lambda) y^j \tag{7.4}$$

wherever (x^i, y^i) and (x^j, y^j) represent two examples of the training data, and $\lambda \in [0; 1]$. Thus, Mixup theory expands the training set via a difficult restriction that linear interpolations of input instances have to result in the straight interpolations of their annotations.

7.7.1 Interpolation Consistency Training (ICT)

ICT method [34] presented to standardizes semi-supervised learning via promoting the model expectation at an interpolation of pair of unannotated samples to be compatible with the interpolation of the network expectations of these instances. Figure 7.7 shows the architecture of the ICT framework. The low-intensity discon-nection hypothesis motivates this approach, and Mixup could accomplish extensive space decision frontiers. Specifically, this can be realized by optimizing the deep network $f(\theta)$ to predict $Mix_\lambda(\hat{y}^j; \hat{y}^j)$ for the paired sample $Mix_\lambda(x^j; x^k)$ using the Mixup strategy: $Mix_\lambda(a; b) = \lambda a + (1 - \lambda)b$. In semi-supervised scenario, ICT improve over Mixup by learning the deep network $f(\theta, x)$ to calculate the "false label" $Mix_\lambda(f(\theta, x^j); f(\theta, x^k))$ for the beforementioned pair of samples. Besides, the network f_θ compute the false label $Mix_\lambda(f_{\theta'}(x^i); f_{\theta'}(x^j))$ for the pair $Mix_\lambda(x^i; x^j)$, where θ' denote the EMA of θ for a further improved consistency regularization (Fig. 7.8).

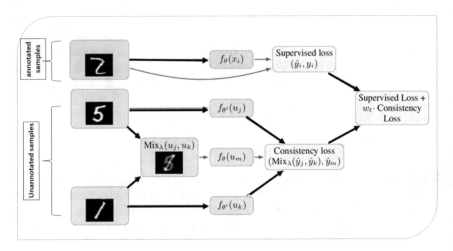

Fig. 7.7 The architecture of ICT framework

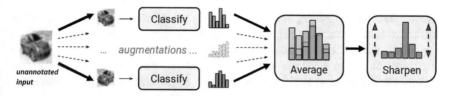

Fig. 7.8 The architecture of MixMatch approach

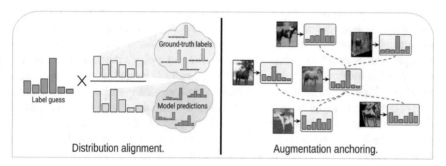

Fig. 7.9 The architecture of ReMixMatch approach

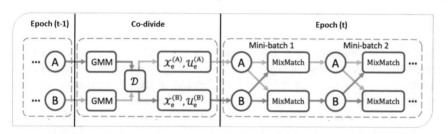

Fig. 7.10 The architecture of DivideMatch approach

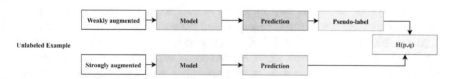

Fig. 7.11 The architecture of FixMatch approach

7.7.2 MixMatch

MixMatch approach [35] integrates entropy minimization and consistency regularization to develop a cohesive cost function for semi-supervised training. The MixMatch conducts by generating pseudo-labels for each unannotated example and later performs training using original annotated data together with the pseudo-labels for the unannotated samples utilizing completely supervised methods. The key objective of this approach is to establish two compilations x_L' and x_U', which are contain augmentation of annotated and unannotated examples that are engendered utilizing Mixup. Officially, MixMatch provide a single augmentation for every annotated example $(x^i; y^i)$ and K softly augmented editions of each unannotated example x^j. Later, it engenders a pseudo label $\hat{y}^j$ for each x^j through calculating the average expectation throughout the K intensifications. The distribution of quasi-label is later refined by adapting temperature scrambling to obtain the ultimate quasi-label. Following the augmentation process, the bunches of augmented annotated samples and quasi-unannotated samples are blended, then the entire set is reordered. This set is partitioned into two portions: the earliest L examples were considered as W^L, while the left over considered as W^U. The company W^L and the augmented annotated batch is passed to the Mixup method to calculate examples $(x'; y')$. In the same way, Mixup is employed between the left over W^U and the augmented unannotated samples. MixMatch achieves conventional fully-supervised learning based on cross-entropy loss for supervised training data, whilst the mean square error is employed for unannotated data provided those mixed-up instances.

7.7.3 ReMixMatch

ReMixMatch [36] is introduced as an extension to MixMatch by aligning the distribution ent and performing an augmentation anchoring. The task of distribution alignment promotes the minimal distribution of accumulated category estimates of unannotated data near to the negligible distribution of the mask. the consistency regularization module of MixMatch was substituted with Augmentation anchoring. This method produces several robustly augmented editions of input and incites every yield to be similar to the expectation from an unconvincingly augmented variation of the identical input. A variation of AutoAugment [37] called "CTAugment" was presented to generate robust augmentations by optimizing and learning an augmentation policy together with the semi-supervised network learning. As the practice of ReMixmatch introduced in Fig. 7.9, an "anchor" is engendered via the application of soft augmentation to a provided unannotated data sample and then use a number of robustly-augmented editions of the same unannotated samples utilizing CTAugment.

7.7.4 DivideMix

$DivideMix$ [38] was introduced as an improved semi-supervised approach to tackle the challenge of learning with noisy annotations. DivideMix is intended to co-divide, which is a procedure that enables two deep networks to learn at the same time. For each of them, an energetic Gauss Mixed Model (GMM) is tailored for the failure distribution of input instances to split the dataset into annotated and unannotated subsets. The divided training sets are subsequently employed for training both deep networks for the following epoch. In the follow-up semi-supervised training, cooperative improvement and cooperative predicting are employed to enhance the ability of MixMatch to learn from noisy annotations.

7.7.5   FixMatch

FixMatch [39] integrates pseudo-labeling with consistency regularization into a single framework, while massively make the overall framework simpler. The main improvement stem from the recipe of these two approaches, and the utilization of a discrete soft and robust augmentation for consistency regularization. Provided a single data example, only at what time the network calculates a great-confidence target, then the generated quasi-annotation might have being pinpointed as label. provided an example x^j, FixMatch primarily engenders pseudo-label $\hat{y}^j$ for inadequately augmented unannotated example $\hat{x}^j$. Later, the network trained with the inadequately augmented instances is exploited to expect pseudo-label in case of the robustly augmented variant x^j. In FixMatch, inadequately augmentation consists of regular flip-and-shift transformations, arbitrarily horizontal flipping of images based on some probability. In a robust augmentation scenario, a couple of augmentation strategies are employed founded on RandAugment [40] and CTAugment [36]. While the Cutout [41] is pursued by one or the other of those strategies (Fig. 7.10 and 7.11).

7.8   Summary and Learnt Lessons

To sum up, the discussion in this chapter taxonomize semi-supervised deep learning into different categories, and offers a detailed discussion of each category that give valuable insights about each category as follow:

- The fundamental concept of consistency regularization approaches revolves that the outcome of the deep network stays unbothered in case of sensible perturbation. Consistency restraints may be set in any of three stages namely neural networks, input dataset, and training operations. From an input standpoint, perturbations are typically injected into the data samples via additive/multiplicative noise, randomized augmentation, or even adversarial training.
- A variety of generative networks can provide efficient semi-supervised solutions by combining unsupervised and supervised loss as single objective functions. The presented GANs approaches follow the standard GAN architecture but exhibit key discrepancies in the amount and structure of the building blocks such as discriminator and generator, etc. these approaches expand the baseline Generative

network through leveraging extra knowledge during the learning i.e., category information, local information, context information, etc.

- The semi-supervised autoencoder applies different versions of autoencoders to learn from labeled and unlabeled data samples. The autoencoder approaches have the advantage of learning insightful representations from data by generative latent-feature networks.
- The major goal of graph-based networks for semi-supervised learning is to carry out label inference on a structured resemblance graph with the aim of enabling the label information to be disseminated from the annotated examples to the unannotated samples by integrating the feature and topologic representation. Furthermore, the commitment of semi-supervised solutions assists in generating further discriminatory embedding representations valuable for the downstream tasks.
- It is worth noting that the hybrid approaches show great success among the semi-supervised deep learning counterparts. This is because they are adapted to tackle the main issue of other semi-supervised approaches hence achieving state-of-the-art performance while improving the suitability for real-world application with large datasets.

References

1. G.J. Qi, J. Luo, "Small data challenges in big data era: a survey of recent progress on unsupervised and semi-supervised methods." IEEE Trans. Pattern Anal. Mach. Intell. (2020). https://doi.org/10.1109/tpami.2020.3031898
2. O. Chapelle, B. Scholkopf, A. Zien, *Semi-supervised Learning* (Massachusettes MIT Press View Artic, Cambridge, 2006)
3. Z. Hailat, A. Komarichev, X. W. Chen, "Deep Semi-supervised Learning," (2018). https://doi.org/10.1109/ICPR.2018.8546327
4. F. Zhuang et al., "A comprehensive survey on transfer learning." Proceedings of the IEEE. (2021). https://doi.org/10.1109/JPROC.2020.3004555
5. D. Zhang, J. Han, G. Cheng, M.H. Yang, "Weakly supervised object localization and detection: a survey." IEEE Trans. Pattern Anal. Mach. Intell., (2021). https://doi.org/10.1109/TPAMI.2021.3074313
6. T. M. Hospedales, A. Antoniou, P. Micaelli, A.J. Storkey, "Meta-learning in neural networks: a survey." IEEE Trans. Pattern Anal. Mach. Intell., (2021). https://doi.org/10.1109/TPAMI.2021.3079209
7. A. Oliver, A. Odena, C. Raffel, E.D. Cubuk, I.J. Goodfellow, "Realistic evaluation of deep semi-supervised learning algorithms," (2018)
8. M. Belkin, P. Niyogi, "Laplacian eigenmaps and spectral techniques for embedding and clustering," (2002). https://doi.org/10.7551/mitpress/1120.003.0080
9. A. Rasmus, H. Valpola, M. Honkala, M. Berglund, T. Raiko, "Semi-supervised learning with Ladder networks," (2015)
10. M. Sajjadi, M. Javanmardi, T. Tasdizen, "Regularization with stochastic transformations and perturbations for deep semi-supervised learning," (2016)
11. S. Laine, T. Aila, "Temporal ensembling for semi-supervised learning," (2017)
12. A. Tarvainen, H. Valpola, "Mean teachers are better role models: weight-averaged consistency targets improve semi-supervised deep learning results," (2017)

13. Z. Ke, D. Wang, Q. Yan, J. Ren, R. Lau, "Dual student: breaking the limits of the teacher in semi-supervised learning," (2019). https://doi.org/10.1109/ICCV.2019.00683

14. H. Navidan et al., "Generative Adversarial Networks (GANs) in networking: A comprehensive survey & evaluation," *Comput. Networks*, 2021, doi: https://doi.org/10.1016/j.comnet.2021.108149

15. J. T. Springenberg, "Unsupervised and Semi-supervised Learning with Categorical Generative Adversarial Networks," (Nov 2015), [Online]. Available: http://arxiv.org/abs/1511.06390

16. E. Denton, S. Gross,, R. Fergus, "Semi-supervised Learning with Context-Conditional Generative Adversarial Networks," (Nov 2016), [Online]. Available: http://arxiv.org/abs/1611.06430

17. Z. Dai, Z. Yang, F. Yang, W. W. Cohen, R. Salakhutdinov, "Good semi-supervised learning that requires a bad GAN," (2017)

18. D.P. Kingma, M. Welling, "Auto-encoding variational bayes," (2014)

19. D.J. Rezende, S. Mohamed, D. Wierstra, "Stochastic backpropagation and approximate inference in deep generative models," (2014)

20. D. P. Kingma, D. J. Rezende, S. Mohamed, M. Welling, "Semi-supervised learning with deep generative models," (2014)

21. M.E. Abbasnejad, A. Dick, A. Van Den Hengel, "Infinite variational autoencoder for semi-supervised learning," (2017). https://doi.org/10.1109/CVPR.2017.90

22. N. Siddharth et al., "Learning disentangled representations with semi-supervised deep generative models," (2017)

23. J. Schulman, N. Heess, T. Weber, P. Abbeel, "Gradient estimation using stochastic computation graphs," (2015)

24. Z. Wu, S. Pan, F. Chen, G. Long, C. Zhang, P.S. Yu, "A comprehensive survey on graph neural networks." IEEE Trans. Neural Networks Learn. Syst., (2021). https://doi.org/10.1109/TNNLS.2020.2978386

25. J Gilmer, S.S. Schoenholz, P.F. Riley, O. Vinyals, G.E. Dahl, "Neural message passing for quantum chemistry," (2017)

26. F. Scarselli, M. Gori, A.C. Tsoi, M. Hagenbuchner, G. Monfardini, "The graph neural network model." IEEE Trans. Neural Networks (2009). https://doi.org/10.1109/TNN.2008.2005605

27. S. Zhang, H. Tong, J. Xu, R. Maciejewski, "Graph convolutional networks: a comprehensive review." Comput. Soc. Networks, (2019). https://doi.org/10.1186/s40649-019-0069-y

28. D. Bahdanau, K.H. Cho, Y. Bengio, "Neural machine translation by jointly learning to align and translate," (2015)

29. Z. H. Zhou, M. Li, "Semi-supervised learning by disagreement." Knowl. Inf. Syst., (2010). https://doi.org/10.1007/s10115-009-0209-z

30. D.D. Chen, W. Wang, W. Gao, Z. H. Zhou, "Tri-net for semi-supervised deep learning," (2018). https://doi.org/10.24963/ijcai.2018/278

31. S. Qiao, W. Shen, Z. Zhang, B. Wang, A. Yuille, "Deep co-training for semi-supervised image recognition," (2018). https://doi.org/10.1007/978-3-030-01267-0_9

32. Z.H. Zhou, M. Li, "Semisupervised regression with cotraining-style algorithms." IEEE Trans. Knowl. Data Eng., (2007). https://doi.org/10.1109/TKDE.2007.190644

33. H. Zhang, M. Cisse, Y. N. Dauphin, D. Lopez-Paz, "MixUp: Beyond empirical risk minimization," (2018)

34. V. Verma, A. Lamb, J. Kannala, Y. Bengio, D. Lopez-Paz, "Interpolation consistency training for semi-supervised learning," (2019). https://doi.org/10.24963/ijcai.2019/504

35. D. Berthelot, N. Carlini, I. Goodfellow, A. Oliver, N. Papernot, C. Raffel, "MixMatch: a holistic approach to semi-supervised learning," (2019)

36. D. Berthelot et al., "ReMixMatch: Semi-supervised Learning with Distribution Alignment and Augmentation Anchoring," (Nov 2019) [Online]. Available: http://arxiv.org/abs/1911.09785

37. E.D. Cubuk, B. Zoph, D. Mane, V. Vasudevan, Q.V. Le, "Autoaugment: learning augmentation strategies from data," (2019). https://doi.org/10.1109/CVPR.2019.00020

38. J. Li, R. Socher, S.C.H. Hoi, "DivideMix: Learning with Noisy Labels as Semi-supervised Learning," (Feb 2020) [Online]. Available: http://arxiv.org/abs/2002.07394

39. K. Sohn et al., "FixMatch: simplifying semi-supervised learning with consistency and confidence," (2020)
40. E.D. Cubuk, B. Zoph, J. Shlens, Q.V. Le, "RandAugment: practical automated data augmentation with a reduced search space," (2020)
41. T. DeVries, G.W. Taylor, "Improved Regularization of Convolutional Neural Networks with Cutout," (Aug 2017) [Online]. Available: http://arxiv.org/abs/1708.04552

Chapter 8
Deep Reinforcement Learning for Secure Internet of Things

Reinforcement learning (RL) is identified as a branch of artificial intelligence (AI) the seek to addresses the dilemma of automated learning of ideal determinations throughout time, which is a popular and broad challenge explored in lots of technical and industrial disciplines. RL nativey integrates an additional dimension (which is typically time, however not of necessity) into learning process, which lays it considerably near to the social awareness of AI.

In dynamic environment, problems that seem static input–output dilemmas come to be dynamic in a broader standpoint. As an example, given a binary supervised classification problem where inputs are labeled normal or anomalous. The training data is prepared passed to the classifier for training purpose, and after some time, the classifier get converged demonstrating robust performance, then deployed keept running for a period. Later, you find out that the patterns of normal data significantly changed, and a noteworthy percentage of input queries are currently misclassified, hence, it is essential to upgrade your training images and replicate the procedure once again. Great?

Absolutely not!

The previous example is aimed to demonstrate that even straightforward AI problems (supervised, unsupervised, semi-supervised) come up with a concealed time dimension, which is commonly ignored, however, it may possibly turn into a problem in the production phase.

In view of this, RL brings up ambient intelligence to the Internet of Things (IoT) by offering a category of approaches to solve the dynamicity dilemma in handling the IoT information to make reactive decisions. In Particular, optimum mapping from states to activities is determined by the learned policies from the interaction between the agents and the enclosing environment. The learning agent ought to

Electronic Supplementary Material The online version of this chapter (https://doi.org/10.1007/978-3-030-89025-4_8) contains supplementary material, which is available to authorized users.

be capable to perceive the present status of the system in some way that enables performing some action that influences the new state and the instant reward with the aim of maximizing the long-standing reward over a prolonged time. Unlike the previous categories of deep learning, the model is not informed about the activities to be performed, however should determine the activities that generate the extremely long-standing reward based on trials. Meanwhile, RL solutions have been magnificently applied to different fields, it faces a challenging task once undertaking issues with practical complications, i.e., the agents should professionally symbolize the environmental status from multi-dimension IoT data and exploit this knowledge to learn optimum decision strategies. Thus, deep learning is exploited to assist the RL to deal with high dimensionality leading to the concept of deep reinforcement learning (DRL).

In the nutshell, this chapter intends to provide a tutorial for Deep Reinforcement Learning that can be exploited to develop a security solution in an IoT-based system. This is discussed through the following sections:

- Foundations and Preliminaries
- Single-agent Reinforcement Learning
- Multi-agent Reinforcement Learning
- Taxonomy of Deep Reinforcement Learning
- Reinforcement Learning based IoT Applications
- Summary and Learnt Lessons.

8.1 Foundations and Preliminaries

It worth noting that developing DRL solution for realistic IoT encvironment s not a simple task as believed. Two main entities are essential for any DRL model, which are the agent and environment. Originally, in DRL, the environment could be constrained to just signify the perception components or be expanded to involve the networking technologies, the edge nodes, fog, or cloud backends. the computational performance, such as energy consumption, communication overhead, and latency, typically have significant consequences in the management efficiency of the physical IoT devices. Hence, in DRL, the actions could be apportioned into two stages of control namely physical control which considers actuator-related actions as well as resources control which considers actions related to energy, computation, and communications [1]. Couple of stages of management could be divorced or cooperatively optimized and learned. In addition, in DRL, the agent represent a rational model for deciding the action assortment. For IoT-based systems, an intelligent agent could be hosted by edge devices, network devices, fog servers, and cloud servers. The time-sensitivity of the system becomes an imperative aspect for regulating the agents' positions. Images of captured by autonomous smart vehicle require to be handled immediately to evade any collision. In this scenario, the vehicle should host the agent locally take decisions rapidly, rather than transmitting the information to a distant server and restoring the decisions backward to the vehicle. Nevertheless, multiple scenarios exist where it is difficult to ascertain the ideal sites for the agents, which could entail resolving an RL dilemma in itself. Furthermore, in the case of many agents disseminated over IoT entities, making the collaboration among agents an essential and interesting matter [1].

8.2 Single-Agent Reinforcement Learning

8.2.1 Markov Decision Process

In reinforcement learning, single intelligent agent operate and interact with elements of IoT ecosystem (environment) to resolve a subsequent challenge encountered in making decision. Completely discernible environment is usually designated as Markov Decision Process (MDPs) that enable the agent to gain entrance to the correct IoT ecosystem's state in every time phase. Provided the system state at a particular time, the agent will make a decision to perform some activities the change the system state into a different state experimented as of probability distribution. The agent receives a reward for its instantaneous return. This is commonly known as an infinite-horizon inexpensive return. An Additional standard design is undis-counted finite horizon return computed throughout a predetermined perspective. That is popular for occasional missions (i.e., jobs with an ending point). The agency intends to discover an optimum policy for mapping between actions and system states while maximizing the anticipated return. A strategy or a tactic explains the agent's behavior at each moment. A policy sends activities for execution in every apparent state. Conversely, a stochastic policy yields dissemination across actions. For any policy, a Q-function can be defined to calculates the estimated collected rewards gazing from any provided state [1] (Fig. 8.1).

Fig. 8.1 Deep reinforcement learning in IoT

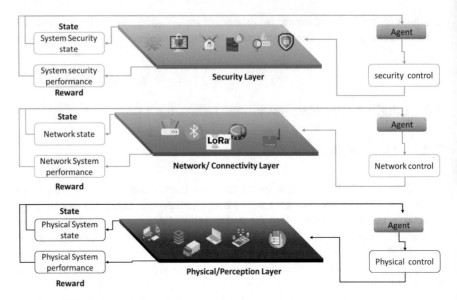

Fig. 8.2 General deep reinforcement learning for internet of things layers

8.2.2 *Partially Observable Markov Decision Process*

Through the earlier discussions, it is believed that the complete state information can be accessed by the agent. Nevertheless, this hypothesis comes to be breached in the majority of actual IoT environments. A typical example for this, IoT entities bring together information regarding the surroundings utilizing sensors. The device weights are boisterous and constrained; therefore, the agent would only possess a piece of deficient information concerning the environment. Numerous issues like anonymity which avert the agent from realizing the complete state knowledge utilizing the sensory data. Thus, Partially Observable Markov Decision Processes (POMDP) presented to offer a generalization of the MDP approach to consider the ambiguity in the state information [2] (Fig. 8.2).

8.3 **Multi-agent Reinforcement Learning**

Multi-agent Reinforcement Learning (MARL) effectively handles the challenges of making serial decisions including a group of agents. Thus, the changing aspects of IoT systems are affected by the multiparty action performed by the involved agents. instinctively, the value of reward accepted from a particular agent becomes an operation of all the agents' activities rather than a function of its individual actions. Thus, to extend the prolonged reward, the agent ought to consider the other agents' policies [2].

8.3.1 Markov/Stochastic Games

Markov Games (MGs) or Stochastic Games (SGs) is considered as an extension of the MDP paradigm to the multiagent configuration to consider the interrelation among different agents [3]. Given the environment including multiple agents, then, at any state, each agent chooses a particular action and the shared action of all would be carried out in the environment, where the transition from one state to another is administrated by the evolution probability function. After that, each agent would be given an instantaneous incentive characterized using a particular reward function [4]. in MG It is worth noting that the value of reward and shifting operatins are reliant on the mutual action space, where every agent aim to discover the ideal policy that boosts its prolonged return. Therefore, the ideal policy of any agent can be defined as a function of the policies of its adversaries. In this way, the complexity of Multi-agent Reinforcement Learning systems evolves from this estate for the reason that the policies of other agents are dynamic and adjust throughout the learning process. In view of this, three solution models can be differentiated for MGs including wholly cooperative, wholly competitive, and integrated. In the former scenario, overall agents share a particular function of reward and thus share identical state–action function. Wholly cooperative MG is also known as Multi-agent MDP (MMDP). Which in turn shortens the dilemma as the typical sole-agent approaches could be employed only when overall agents are organized utilizing a main entity. Conversely, wholly competitive MG is handled through investigating for a Nash Equilibrium [5] (Fig. 8.3).

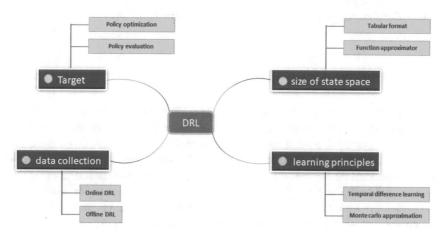

Fig. 8.3 Classification of diverse RL sceneries

8.3.2   Dec-POMDP

in the multi-agent scenario, increasing the number of agents includes extra intricacy in discovering the agents' optimum policies. As a matter of fact, every agent requires complete knowledge regarding the actions of other agents for capitalizing the coressponding values of rewards. Thus, the ambiguity regarding the adversaries along with the state ambiguity necessitates an expansion of the MG structure to handle collaborative agents in case of incomplete discernable environments. In view of this, the Decentralized POMDP (Dec-POMDP) is the approved numerical solution for examining the collaborative progressive decision related troubles in case of ambiguous conditions. That can be considered as an explicit generalization of POMDPs for the multi-agent scenario [2].

A Dec-POMDP is a particular circumstance of the Partially Observable Stochastic Games (POSG) in which all the components are similar to Dec-POMDP with an exception for the reward function that turns out to be specific toward every agent. POSG empowers the development of autonomously attracted agents although Dec-POMDP completely provides collaborative agents under incomplete discernable conditions. every agent obtains the corresponding observation with no realization of observations of other agents for every particular state. Therefore, each agent selects an action that results in a cooperative action achieved in the environment. Founded on a shared instantaneous reward, each agent attempts to discover a regional policy for amplifying the longstanding reward of its squad. The policy is termed local for the reason that each agent functions in accordance with its personal local interpretations without transmitting or distributing information with the others [3].

8.3.3   Networked Markov Games

Collaborative Dec-POMDPs or MGs are only appropriate for standardized collaborative agents as they have the same reward indicator. Nevertheless, the majority of practical IoT ecosystems include dissimilar agents having different inclinations and objectives. Additionally, possessing an identical reward function necessitates public knowledge as of all the agents for approximating a public state–action function that obscure the federalization of this approach. To tackle those issues, Networked MG simplifies the MG design to train collaborative agents based on various incentive functions through exploiting mutual knowledge throughout an IoT ecosystem.

In the nutshell, the benefits of interacted MGs in comparison with traditional MGs are (1) the opportunity to model diverse agents having various reward functions; (2) declining the synchronization charge by bearing in mind national-to-national communication which simplifies the design of distributed multi-agent DRL; (3) the privacy-protecting estate as agents are not authorized to reveal their reward functions [5].

8.4 Taxonomy of Deep Reinforcement Learning

To satisfy the various requirements of dissimilar IoT ecosystems, variations of DRL approaches have been evolving throughout the ages. According to the target (i.e., policy or reward) of the RL algorithm, the current DRL approaches could be divided into policy-based and value-based.

8.4.1 Value-Based DRL

In this value-centered approaches, as a replacement for depositing an unambiguous policy, the value function (VF) is estimated using the DRL algorithm, by approximation of the anticipated prolonged state's reward. Therefore, the VF could be described as a relation between a given state and the related prolonged reward [6]. Besides, drawn from state–value function, the Q-function or action–value function is established for mapping of certain state–action couples to prolonged values of reward. The Q-function function is especially valuable for concentrating a specific activity within a certain system state. state–value function varies from Q-function based on the fact that preliminary action is provided in the Q-function. Besides, in value-centered approaches, autonomous agents seek for acquiring the ideal manage policy that surpasses other strategies in the collective reduced values of reward. Hence, the challenging of finding the ideal policy is altered into the challenge of approaching the ideal Q-function. In this regard, two typical strategies to accomplish that, which are Temporal Difference (TD) and Monte Carlo (MC) approximation [7].

Firstly, MC approximation regularly renovates the Q-function and enhances the strategy based on the assumption of comprehensive strategy reiteration. Specifically, the Q-function is upgraded according to the averaging prolonged rewards of ancient knowledge. With enough knowledge, the Q-function will approximate the genuine Q-function in regard to the existing strategy. Given that hypothesis, MC strategy advances the policy via the creation of a modern policy that is selfish regarding the existing Q-function. In Particular, for all states, the fresh selfish policy performs the action with the greatest prolonged reward in a deterministic manner.

Secondly, TD learning pursues an analogous iterative learning procedure, while excluding the upgrade rule of Q-function. It utilizes the concept of Dynamic Programming to grasp the partial installments through bootstrapping. In other words, the intermediary reward and the expected cost of the upcoming system state are employed to upgrade the Q-function [8].

8.4.2 Policy-Based DRL

This subsection presents policy-based DRL approaches. as previously stated, the value function is employed to deduce the policy of the value-centered RL algorithm. On the other hand, policy-centered approaches aim to model the policy and unambiguously utilize parameter-based estimators (such as deep learning models) to set the policies. In this regard, two major categories of policies. First, deterministic policy, which performs direct mapping of each state to a particular action. Second, the stochastic policy can be employed to perform a mapping from a particular state to the probability distribution throughout the action space. Bearing in mind, the deterministic policy could be construed as an indiscriminate policy by setting 1 as the possibility of the objective action while setting 0 as the probability of the outstanding activities [9].

8.5 Reinforcement Learning Based IoT Applications

8.5.1 IoT-Based Industrial IoT

The extensive approval of IoT technologies in the Industry 4.0 imposes elevated conditions on system constancy and security. The shortage of efficient defense or resistance is the probable obstruction for upcoming IIoT ecosystem. Mercifully, as a result of introducing artificial intelligence, DRL approaches have demonstrated wonderful accomplishment in handling complicated challenges without an earlier understanding of the system. In view of this, preventive maintenance is identified as a reduction of the chance of equipment malfunction via routine equipment maintenance, i.e., machine substitute and adjustment. Inspired by this fact, Q-learning can provide a founding solution for preventative equipment substitution [10].

Generally, machine state in IIoT systems worsens over time, which as a result lead to malfunctions of equipment. It is a problem that the expense of frequent machine substitutions might overshadow the commercial profits, whilst ineffective substitutions could improve the possibility of malfunction. Consequently, the aim of Q-learning is to decide the best substitute timing. The state is characterized by the decline of machine performance. This way, the Q-learning overcomes the heuristic approaches as it is susceptible to the smallest substitution periods devoid of endangering the functioning steadiness during the simulative experimetns. In another way, the Monte-Carlo RL algorithm can be employed to extent the conventional DRL methods to troubles having a larger search domain, which explicitly learns from the mean sampling yields of computer-generated practices. Besides, POMDP can also be employed for preventive maintenance dilemmas, where a completely circulated multi-agent deep Q-learning method can be used to handle partial observability, where non-positive value is set as the incentive of every autonomous agent and computed as the total system cost [11].

On the other hand, cyber-physical security can be defined as a set of techniques and policies for maintaining the security of the devices, data, network, and control systems from incidents or illegal entrance. The reliance of industrialized mechanization procedures on the aggregation and distribution of data across IoT entities make them prone to physical and/or cyber outbreaks. This issue can be designated as as a noncollaborative game that can be solved by a multi-agent approach (i.e., Q-learning). the invaders try to maneuver the communicated data by fictitious data injection as well as intelligent jamming with the main to provoke equipment crashes [12].

Instead, the automatic system seeks to organize the mutually dependent equipment aiming to guarantee the reliability of the IIoT system. Contemplating the system's unnoticeability, the agents in DRL can be installed to autonomously manage the mechanized procedure and invaders. Moreover, agents in deep Q-learning can operate like a valuable mechanism for detecting nasty attacks through data transmission. Nevertheless, thanks to the fact that some security dilemmas do not have pre-determined measurements, some studies have to apply some unintended assessments which might be erroneous. future research is anticipated to develop an efficient evaluation measure to address this problem [13].

8.5.2 IoT-Based Intelligent Transportation

The integration of wireless IoT sensors, improved communications, and adaptive management schemes in an Intelligent Transportation System (ITS) enable realizing a more secure, more organized, and effective transportation network. In view of this, this section introduces and discusses the typical applications of DRL in IoT-enabled ITS.

For Vehicular Ad hoc Networks (VANETs), security attacks such as spoofing, eavesdropping, and DDoS could result in packet defeat, vulnerable communications, system failure, or even disasters [14]. Taking into account the multiplicity of attack patterns and the sovereignty of attackers, some efforts have been devoted to improving communication security based on adjustive energy management. The system was framed like a noncollaborative game between malevolent attackers and transmitters. To this end, agents in Q-learning can be employed for assigning the transmission energy throughout various stations without knowing the channel or attacker models. The simulation results of this solution show that the potentials of attackers are greatly restricted in the game, in that way destroying the attack. Motivated by this effort, an intelligent deep Q-learning-centered energy distribution procedure is designed by exploiting the receiver filter and beamforming s against spoofing and eavesdropping, respectively. Anti-jamming is believed as one more security challenge in ITS communication, whereas radio noise is introduced to leads to transmission failures intentionally [15]. This is distinct from the non-intentional intervention of radio noise. To this end, the anti-jamming challenges are normally articulated as Partially Observable Markov Decision Processes. In another way, multiple agents can run deep Q-learning to provide anti-jamming energy management via interaction with smart assailants, which in turn enable the vehicular defending agents to define the relay nodes and transfer powers. The state involves the connection traits, signal-to-interference-plus-noise-ratio (SINR), and bit-error-rate (BER). This solution enables great improvement of the network utilization and lessens the messages' BER [16].

8.6   Summary and Learnt Lessons

This chapter provides an in-depth discussion about the role of deep reinforcement learning for securing IoT environments by enabling efficient and effective defense against attacks. Besides, it also provides a taxonomy for classifying the current deep reinforcement learning according to different perspectives. In the nutshell, this chapter delivers useful insights regarding the potential of deep reinforcement learning for maintaining the security of IoT environments:

- This chapter provides an in-depth discussion for different categories of deep reinforcement learning under single-agent as well as multi-agent scenarios. It provides fine-grained taxonomy for deep reinforcement learning algorithms based on its structural design.
- This chapter shows that deep reinforcement learning can be effectively applied for securing and defending IoT-enabled smart grids against a form of cyber-attacks. it worth observing that the simplicity of the Q-learning mechanism gains more considerations compared to other DRL algorithms in smart grid. Nevertheless, the applicability of DRL in such an application scenario are comparatively hypothetical.
- It also demonstrates the effectiveness of deep reinforcement learning in IoT-based Intelligent Transportation systems. it worth noting that recent academic and industrial efforts emphasize the distributed deep reinforcement learning and are more concerned with the multi-agent scenario, which is instigated by scalability problems, latency, data synchronization. numerous challenges continue in developing a multi-agent deep reinforcement learning approach with rapid convergence and steady performance. This is because of the absence of efficient synchronization procedures between distributed agents.

References

1. L. Lei, Y. Tan, K. Zheng, S. Liu, K. Zhang, X. Shen, Deep reinforcement learning for autonomous internet of things: model, applications and challenges. IEEE Commun. Surv. Tutorials (2020). https://doi.org/10.1109/COMST.2020.2988367
2. A. Feriani, E. Hossain, Single and multi-agent deep reinforcement learning for AI-enabled wireless networks: a tutorial. IEEE Commun. Surv. Tutorials (2021). https://doi.org/10.1109/COMST.2021.3063822

3. L. Yu, S. Qin, M. Zhang, C. Shen, T. Jiang, X. Guan, A review of deep reinforcement learning for smart building energy management. IEEE Internet Things J. (2021). https://doi.org/10.1109/JIOT.2021.3078462

4. X. Liu, W. Yu, F. Liang, D. Griffith, N. Golmie, On deep reinforcement learning security for Industrial Internet of Things. Comput. Commun. (2021). https://doi.org/10.1016/j.comcom.2020.12.013

5. N.C. Luong et al., Applications of deep reinforcement learning in communications and networking: a survey. IEEE Commun. Surv. Tutorials (2019). https://doi.org/10.1109/COMST.2019.2916583

6. L. Qian, Y. Wu, F. Jiang, N. Yu, W. Lu, B. Lin, NOMA assisted multi-task multi-access mobile edge computing via deep reinforcement learning for Industrial Internet of Things. IEEE Trans. Ind. Inform. (2021). https://doi.org/10.1109/TII.2020.3001355

7. H. Yang, W. De Zhong, C. Chen, A. Alphones, X. Xie, Deep-reinforcement-learning-based energy-efficient resource management for social and cognitive internet of things. IEEE Internet Things J. (2020). https://doi.org/10.1109/JIOT.2020.2980586

8. Y. Wu, Z. Wang, Y. Ma, V.C.M. Leung, Deep reinforcement learning for blockchain in industrial IoT: a survey. Comput. Netw. (2021). https://doi.org/10.1016/j.comnet.2021.108004

9. A. Uprety, D.B. Rawat, Reinforcement learning for IoT security: a comprehensive survey. IEEE Internet Things J. (2021). https://doi.org/10.1109/JIOT.2020.3040957

10. M.S. Frikha, S.M. Gammar, A. Lahmadi, L. Andrey, Reinforcement and deep reinforcement learning for wireless Internet of Things: a survey. Comput. Commun. (2021). https://doi.org/10.1016/j.comcom.2021.07.014

11. X. Wang, C. Wang, X. Li, V.C.M. Leung, T. Taleb, Federated deep reinforcement learning for Internet of Things with decentralized cooperative edge caching. IEEE Internet Things J. (2020). https://doi.org/10.1109/JIOT.2020.2986803

12. H.A. Shah, L. Zhao, Multiagent deep-reinforcement-learning-based virtual resource allocation through network function virtualization in Internet of Things. IEEE Internet Things J. (2021). https://doi.org/10.1109/JIOT.2020.3022572

13. K. Li, W. Ni, E. Tovar, M. Guizani, Joint flight cruise control and data collection in UAV-aided Internet of Things: an onboard deep reinforcement learning approach. IEEE Internet Things J. (2021). https://doi.org/10.1109/JIOT.2020.3019186

14. G. Mei, N. Xu, J. Qin, B. Wang, P. Qi, A survey of Internet of Things (IoT) for Geohazard prevention: applications, technologies, and challenges. IEEE Internet Things J. (2020). https://doi.org/10.1109/JIOT.2019.2952593

15. H. Xu, X. Liu, W.G. Hatcher, G. Xu, W. Liao, W. Yu, Priority-aware reinforcement-learning-based integrated design of networking and control for Industrial Internet of Things. IEEE Internet Things J. (2021). https://doi.org/10.1109/JIOT.2020.3027506

16. L. Yang, M. Li, P. Si, R. Yang, E. Sun, Y. Zhang, Energy-efficient resource allocation for blockchain-enabled Industrial Internet of Things with deep reinforcement learning. IEEE Internet Things J. (2021). https://doi.org/10.1109/JIOT.2020.3030646

Chapter 9
Federated Learning
for Privacy-Preserving Internet of Things

The rapid evolution of the Internet of Things (IoT) and relevant applications have been paving the way toward the fulfillment of smart cities. Smart cities are thought to come up with multiple crucial smart IoT applications i.e., smart manufacturing, smart transportation, autonomous driving/flight, smart buildings, smart healthcare, smart grid, and so many others [1]. Effective deployment of these IoT applications and services and making them available for users usually entail a vast number of IoT devices. Conferring to some global statistics, the total of IoT devices is believed to rises to 125 billion by 2030. This in turn reflects expected large growth in the amount of IoT-generated data. some other statistics show that, by 2025, the size of IoT-generated data will grow up to reach 79.4 zettabytes (ZB). That incredible growth in the size of IoT networks and associated data volume afford fascinating opportunities for deep learning in IoT environments. To this end, one can use centralized deep learning frameworks (discussed in Chaps. 4–7) to support the design of smart IoT applications. Nevertheless, centralized deep learning solutions struggle with the intrinsic problem of leakage of privacy of users owing to the requirement to transmit the local data at IoT devices data to a central third-party server (i.e., cloud server or datacenter) for training purposes. Additionally, centralized deep learning could not be viable in the case of big volumes of data (e.g., astronomical data) and placed at various geographic locations.

To tackle the above-mentioned challenge of big and geographically dispersed data, distributed deep learning solution emerged to disseminate learning workload across manifold computing agents or devices. Those manifold devices could later train many models at distributed locations simultaneously to construct a cohesive deep learning model [2]. Distributed learning empowers more efficient computing through concurrent training of deep learning architecture. Two types of learning approaches, namely data-parallel and model-parallel approaches, could be exploited

Electronic Supplementary Material The online version of this chapter (https://doi.org/10.1007/978-3-030-89025-4_9) contains supplementary material, which is available to authorized users.

M. Abdel-Basset et al., *Deep Learning Techniques for IoT Security and Privacy*,
Studies in Computational Intelligence 997,
https://doi.org/10.1007/978-3-030-89025-4_9

215

to achieve distributed learning solutions. In a data-parallel scenario, the entire data is split into partitions that are later distributed among learning participants for training, whereas each of them is hosting the same model. In the model-parallel scenario, the entire set of data are made available to each participant to train distinctive sections of the learning solution. This may not be appropriate for some IoT applications for the reason that the majority of learning models could not be partitioned into pieces [3].

9.1 Definition of Federated Learning

The notion behind the FL paradigm is to enable different data owners (participants) to collaboratively train a joint ML/DL without revealing their locally stored private data, hence relieving their privacy worries [4]. Generally, the FL framework consists of two main entities namely the participants and the FL server. Each participant trains their local model using the private data and uploads the calculated parameters up to the FL server. Once all participants finish uploading, the FL server aggregate these submitted parameters are aggregated to calculate the global model parameters. Bearing in mind, the standard definition of FL systems presume that the participants are trusted, which means they use their real private data to do the training and submit the true local models to the FL server [5]. this hypothesis cannot forever be credible as discussed in a later section. Contrasted with other mechanisms including centralized learning, FL is able to achieve satisfactory performance with minimal communication burden, while preserving the privacy of participants' data [6]. Table 9.1 overview and compare the FL with different common learning mechanisms.

9.1.1 Federated Training

Typically, the procedure of federated training pass through the following three phases:

Phase 1 (Initialization): The server chooses the target task/application for which the training will be started and determine and declare the relevant data obligations. It also specifies the model's hyperparameters and the training configurations (i.e., weight decay, learning rate, etc.). Subsequently, the initialized model is broadcasted to the active participants for local training.

Phase 2 (Local Training): After obtaining the initial model, every local node starts the learning procedure utilizing the corresponding private samples and iteratively optimizes its local parameters according to a predefined cost function. For each communication round between the participant and the FL server, the updated parameters of each participant are uploaded to the FL server.

Phase 3 (Server-Side Upgrade): By uploading all local parameters, the FL server employs a predefined aggregation method to gather the local information and then

Table 9.1 Comparison of different categories for aggregation methods for federated learning

Aggregation technique	Challenge	Primary notion	Discussion
FedSGD	Statistical	Based on the configurations large-batch synchronous stochastic gradient descent	- Local gradients are shared instead of parameters - Single step of gradient descent is performed per communication round
FedAvg		Participants carry out multiple batch upgrades using the local data to upload the updated parameters to the FL servers instead of local gradients	- Statistically, the FedAvg is demonstrated to diverge under situations where the data is irregularly dispersed among participants - Systematically, FedAvg did not permit participants to execute varying quantities of local learning depending on relevant restraints
FedProx		The local training for each participant incorporates a proximate factor aiming to restrict the contribution of local upgrades to the global model	- Finetune the convergence of models in case of high heterogeneity data - Like FedAvg, the FedProx did not consider the computing power of participants, and thereby treat them evenly during the aggregation
FedMA		Consider the mutation invariance of the neurons prior to completing the aggregation to facilitate adjustment of the size of the final model	- Employ Bayesian non-parametric technique to adapt the size of the global network to the heterogeneousness dissemination of data - The FedMA is susceptible to contaminating attack, in which an opponent could effortlessly trap the federate scheme to enlarge the final network to set up contaminated local network

(continued)

Table 9.1 (continued)

Aggregation technique	Challenge	Primary notion	Discussion
FedPAQ	Communication	The participants are allowed to conduct several local parameter updates before sharing the updated parameters with the server	- FedPAQ compute the global parameters by averaging the local parameters, which necessitates high complication in either greatly convex or non-convex situations
HierFAVG		A tiered participant-edge-cloud aggregation construction in which edge tier aggregate the local updates from the participants, then upload them to the FL server at the cloud	- This multi-level configuration facilitates effective model interchange throughout the present edge-cloud network - It is prone to the troubles of laggards and machine failures
Turbo-Aggregate	Communication and security	A multi-grouping scheme that divides the participants into multiple groups where the parameter upgrades are common between parties in a rotary fashion. A collective trusted communicating method is employed to protect the confidentiality of local data	- It is relatively appropriate for wireless network topology, where network circumstances and user accessibility can change rapidly - The employed secure aggregation method is effective in handling user failures, yet unable to acclimate to new participants joining the network. Thus, it necessitates developing a self-configurable protocol to assist new participants by in such a way that preserve the privacy

sends the updated global model parameters back to the data owners. The server also optimizes a predefined global cost function, that is used to calculate the global parameters. Then, the global parameters are broadcasted back to the participants, where phase 2 and phase 3 are repeated till the global model reach desired performance or convergence happens. Different forms of aggregation methods have been developed to address different FL challenges, as tabulated in Table 9.1.

9.2 Taxonomy of Federated Learning Solutions

This section provides a taxonomy for categorizing federated deep learning according to different facets: splitting methodology, privacy-preservation techniques, interaction scheme, and heterogeneity handling. For convenience, these aspects, the corresponding categories, merits, and applications of those different categories is summarized in Table 9.2.

Table 9.2 Taxonomy of federated deep learning approaches

Aspect	Categories	Advantage	Applications
Splitting methodology	FTL	Upsurge the number of samples and expanding the dimensionality	Knowledge transfer, cross-domain learning
	VFL	Expansion of input dimensionality	Distributed dep learning solutions
	HFL	Rise the number of samples	Learning on resource-constrained devices
Privacy-preservation techniques	Secure parameters aggregation	Escape communicating the initial data	Deep federated solutions
	Homomorphic encryption	Clients could compute and handle the encoded data	Distributed learning
	Differential privacy	Could effectively safeguard the confidentiality of users' data by injecting noise	Privacy preserving artificial intelligence solutions
Heterogeneity handling	Asynchronous communication	Enable handling the communication latency issues	Device heterogeneity
	Sampling	Evade immediate learning with various IoT devices	Pulling reduction with local compensation (PRLC)
	Fault-tolerant approaches	Could thwart the entire system from disintegrating	Laying-off techniques, Sensitive applications
	Heterogeneous design	Could handle the issues related to the heterogeneous devices	5G, B5G, 6G systems

9.2.1   Data Partition

Based on various forms of data distribution in feature space and sample space, as presented in Fig. 9.2, federated learning solution could fall under one out of three main categories namely horizontal federated learning (HFL), vertical federated learning (VFL), and federated transfer learning (FTL) (see Fig. 9.1).

9.2.1.1   Horizontal Federated Learning

HFL is appropriate under a scenario in which the features of two participants' data greatly overlap, but the participants overlap slightly. HFL can be considered as a mean for horizontal dividing of the training data via the participant's dimension, then bring out the portion of the data with the same feature set, but participants remain not precisely the equivalent for learning. In particular, various data samples come up with identical data features (affiliated by features of participants) [7]. Consequently, HFL could rise the sample size of participants. In HFL, it is popular for all participants to evaluate and push up local gradients to a centralized server for aggregation, thereby obtaining the final model parameter. The computation and transmission of gradients information in HFL may disclose participants' sensitive information (see Fig. 9.2).

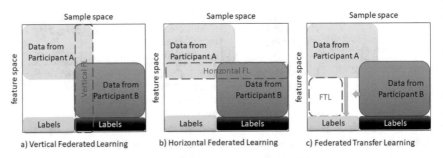

Fig. 9.1 Taxonomy federated learning approaches along with main difference

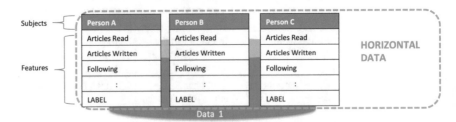

Fig. 9.2 Illustration of horizontal federated learning

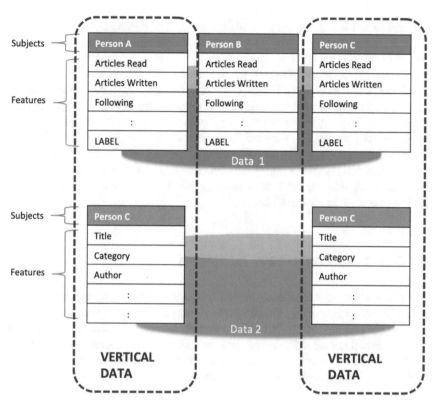

Fig. 9.3 Illustration of vertical federated learning

9.2.1.2 Vertical Federated Learning

VFL is presented when the participants' features of the two or more data exhibit a slight overlap while presenting an excessive participants' overlap [1]. VFL seeks to split the datasets vertically based on feature domain, then wipe out the portion of data with a single set of participants whilst features are not identified during the training. In particular, data from distinct columns have constant participants [5]. Thus, VFL could expand the feature space of the training data. VFL can aggregate these distinct sets of attributes in an encoded form for improving the capability of the learner network. At this time, multiple learning solutions have been demonstrated successful under this kind of federated system [3] (see Fig. 9.3).

9.2.1.3 Federated Transfer Learning

FTL represents the scenario in which the participants and features of the involved data datasets infrequently overlap, hence, no data split is required, but learning is applied to overwhelm the shortage of data or devices [7]. For instance, given two distinct institutes in different countries, thanks to the geographical constraints, the participant groups of the two institutes come up with slight overlap. Simultaneously, owing to the various categories of institutes, the data features of the two datasets are only a small part of the overlap [5]. Hence, realizing an efficient federated solution can be achieved by applying transfer learning to address the troubles of tiny autonomous dataset as well as insignificant annotations, in such a way to enhance the efficacy of the underlying model. The extremely appropriate condition for transfer learning is when you attempt to improve the functioning of a model with insufficient associated training data. FTL enables assuring the confidentiality of participant's data, however, it also transmits the deep network of ancillary tasks to executive training and handles the issues of limited annotated data.

9.2.2 Privacy-Preservation Techniques

The extremely crucial characteristic of FL is that collaborative participants could maintain on the relevant data samples in the vicinity and want to communicate model gradient for training the objective deep network, however, the local parameters are prone to reveal certain confidential data. To this end, three categories of methods can be used to safeguard the federal privacy of the global model namely federated aggregation, homomorphic encryption (HE), and differential privacy (DP) [8].

9.2.2.1 Federated Aggregation

Federated model aggregation can be considered as of the highly popular privacy methods in federated learning solutions that seek to trains the final deep network by briefing and averaging the local gradient from all participants in such a way that evades the transmission of the local data during the training phase. In view of this, some studies attempted multiple IoT devices to contribute to federated training depending on the formed incentive method. To gain effective solutions, the resource efficiency should be immediately optimized throughout the gradient exchange process. Contrasted with creating incentive methods, another methodology shows that local adjustment centered on multi-task learning, refinement, and information mining is a promising means for preserving and enhancing the privacy of local participants, as well as the performance of the federated solution. Consequently, participants can take the advantage of FL and attain improved performance than centrally trained deep learning solution without negotiating the confidentiality of the deep learning model [9].

9.2.2.2   Homomorphic Encryption

Common encryption methods concentrate on securing stored data. It is not possible for participants to access any information without having a key needed to obtain the initial data from the encryption findings, and they are unable to carry out any computation procedures on the encrypted data, or else it would result in failed decryption [1]. Nevertheless, HE could offer a solution to the problem of processing encrypted data, for the reason that it is focused on the security of data handling. The highly critical characteristic of HE is that that the IoT participants could compute and deal with the encrypted data samples while preventing the exposure of the original data. Simultaneously, the participants possessing the key can easily decrypt the secured data as typically expected. HE has become important for federated learning. In particular, during the federated training, the gradient communication between the participants and the aggregation server might leakage the confidential information of participants. HE provides a great solution for this problem by encrypting the parameters of the models before uploading and downloading operations without influencing the performance of the federated model [10].

9.2.2.3   Differential Privacy

Differential Privacy (DP) can be defined as a popular privacy description introduced by 2006 for solving the issues of privacy exposure in statistical databanks. According to this conceptualization, the computation outcomes of certain dataset are not sensitive to the alterations of a particular data instance, and one data record just has a slight influence on the computation outcomes [8]. Thus, the threat of privacy exposure triggered by the accumulation of data records into the database is regulated on an extremely insignificant and tolerable scale, and the attacker is unable to find precise private information by monitoring the computation outcomes [11]. In the training process of conventional learning solutions, it is common to combine noise to the outcome of applying the DP in the procedure of gradient recapitulation to reach the objective of safeguarding participant privacy [12]. Actually, exponential mechanism and Laplace mechanism are normally employed to accomplish DP security. Many research efforts are devoted to exploring two facets of privacy preservation and performance. Injecting extra noise would unavoidably influence the performance. hence, realizing effective equilibrium between privacy and performance has currently become the most prevalent research direction. For instance, DP could be integrated into the model compression method to expand privacy advantages at the same time as enhancing performance. DP can be categorized as global DP and local DP. All the categories of DP techniques could assure $\varepsilon - th$ differential needs of a solo participant, however, the usage settings are somewhat distinct [12].

Differential privacy (DP) is a popular method for privacy protection in artificial intelligence and constitutes a robust typical for privacy assurances for artificial intelligence approaches the learn from the combined dataset. Informally, DP seeks to offer a confine ϵ, that the invaders might learn almost nothing extra about a subject compared to what they learn when it was absent from the dataset since the subject's important data is nearly unrelated to the model's outputs. The confine ϵ signifies the level of privacy predilection that can be managed by various parties. Several studies have tried to maintain DP at the data level in a centralized training scenario. For safeguarding the data from a contrary attack, i.e., use model parameters to deduce the data. To address this limitation, a DP technique for differential privacy protection could be employed during the federated training procedures in IoT [12].

Having a definite mapping method $h : D \rightarrow \mathbb{R}^m$, the sympathy s_h is declared to be the supreme of the absolute difference $h(D) - h(D')_1$, wherever $D - D'_1 = 1$, which represents a single element of variance separating D and D'. Herein, the h calculates the parameters m in the DL architecture. Presenting "noise" through the training procedure (i.e., inputs samples, parameters, or segmentation results) can restrain the granularity of common information and guarantee to realize the DP for all $S \subseteq Range(h)$, and then

$$\Pr[h(D) \in S] \le e^\epsilon \Pr[h(D') \in S], \text{ or} \qquad (9.1)$$

$$\Pr[h(D) \in S] \le e^\epsilon \Pr[h(D') \in S] + \delta \qquad (9.2)$$

where the δ represents the probability that the ϵ-th DP being intruded. At this point, we present two techniques that could realize efficient privacy assurances by incorporating noise into the shared network parameters.

First, the Gaussian technique include the noise $N(0, sd_h^2\sigma^2)$ with average 0 and standard deviation $sd_h\sigma$ to the operation $h(D)$ that has a universal sympathy $s_h.h(D)$ that will follow the DP with (ϵ, δ) when the $\delta \ge \frac{4}{5}\exp(-(\sigma\epsilon)^2/2)$ and the $\epsilon < 1$. Hence, we associated the parameter of Gaussian noise σ to both privacy parameters σ and δ. Second, the Laplace technique, in which the Laplace distribution is positioned around zero and has a scale b that represents the corresponding probability density function as presented in Eq. (9.3).

$$Lap(b) = Lap(x|b) = \frac{1}{2}\exp\left(-\frac{|x|}{b}\right) \qquad (9.3)$$

The Laplace variance is denoted as $\sigma^2 = 2b^2$. This technique includes the noise $Lap(s_h/\epsilon)$ in the function $h(D)$ that has a universal sympathy s_h and conserves DP with $(\epsilon, 0)$. Hence, we associate the parameter b of the Laplace noise to the parameter of the privacy ϵ. In the COVID-19 segmentation scenario, the operation h is implemented using a deep learning segmentation network and it is uncontrollable to calculate the sympathy s_h. For clarification, the sympathy s_h is declared to be one [11].

Exponential Technique: a common DP method that has been suggested for circumstances where injecting noise explicitly to the productivity function (as with the Laplace technique) will absolutely damage the outcomes. Therefore, this technique represents the building element for queries with random value aiming to *maximize* the utility whereas maintaining privacy. Given a random scope S, the mapping operation $h : N^{|X|} \times S \to \mathbb{R}$ take the responsibility of mapping the input–output couples to specific score using the function presented in Eq. (9.4).

$$\Delta h = \max_{s \in S} \max_{D - D'_1 \leq 1} h(D, s) - h(D', s)_1 \qquad (9.4)$$

whereas the sensitivity of h with respect to the input is significant, while it can be arbitrarily sensitive with respect to the scope $s \in S$. This technique $M_E(D, h, S)$ is characterized using a randomized algorithm that selects the output as a component of the scope $s \in S$ using the possibility relative to $\exp \frac{(\epsilon \cdot u(D,s))}{(2\Delta u)}$.

Once normalization is performed, a probability density operation is implemented over the possible solutions $s \in S$. Nonetheless, the resultant distribution might be complicated and span indiscriminately huge realm, hence, the application of this technique might not constantly be efficient. Such a technique can be demonstrated as a $(2\epsilon \Delta u)$-DP technique [10].

In the nutshell, the notion behind DP techniques is combining a specific quantity of noise to the local training parameters, whilst maintaining the value of the initial gradient. This injected noise enables calibrating to the factors (ϵ, δ) of privacy. Therefore, it regulates the parameters to be contingent with privacy obligation by associating the privacy and noise parameters together.

9.2.3 Communication Architecture

The practical application of federated learning is confronted with many issues like unfair data distribution among participants, heterogeneous equipment and systems, computation energy, etc. With the rapid proliferation of IoT devices across different application domains, a variety of not independent and identically distributed (Non-IID) data necessitate to be handled without disclosing confidential information [13]. Based on the real complicated condition, selecting a proper training technique is beneficial to the execution of the deep learning model. In the strategy of decentralized learning, all isolated IoT devices could interact with the centralized server and partake in updating the final federated deep model.

Under a federated scenario, the elasticity of local updates and participant contribution influence the final performance of the model [14]. to this end, FedProx model is presented to bring together edge data for decentralized training and utilizes a federal averaging to guarantee the toughness and permanency of the objective task. Besides, Federated Averaging is a popular federated gradient aggregation algorithm that averages the indiscriminately dwindling gradient information uploaded by federated participants, and then computes the global updates, which are later broadcasted back to local participants [10, 15].

9.2.4   Heterogeneity Handling

The continuously evolving nature of the IoT environment brings us a more complex environment containing a heterogeneous set of types of equipment and devices. In this regard, developing federated learning based on different equipment is commonly known to lead to inefficient performance. To tackle the dilemma heterogeneity, four categories of distraction can be employed namely asynchronous communication, sampling techniques, fault-tolerance method, and learning heterogeneousness [15].

9.2.4.1   Asynchronous Communication

In the conventional design of the data center, two popular communication strategies depending on parallel reiterative optimization methods namely synchronous and asynchronous communication. Nevertheless, considering the variation of IoT devices, the synchronous method can be definitely perturbed, hence, in the federated learning scenario, the asynchronous method is likely to offer an effective solution for the communication dilemma of distributed devices [2]. In this regard, data scarification could provide a solution to the heterogeneity problem during the federated training. Though asynchronous upgrade has been showing great success in distributed and federated systems, however, the conundrum of latency and communication overhead exacerbates the drawback of device heterogeneousness making it the improper choice for time-sensitive applications. to sum up, asynchronous communication can be viewed as the first option to deal with the heterogeneity of participants in the federated system [15].

9.2.4.2   Sampling

In federated learning-based IoT solutions, not all participants want to take part in every training step. In some cases, the participants are nominated to contribute to the federated training, while in an alternative scenario, the participants themselves decide to join in the federated training. More effectively, some researchers consider the development of an incentive method depending on the contract principle, which aims to promote the participants that have high-quality data to enthusiastically contribute to the efficient federated learning with enhanced learning performance.

9.2.4.3   Fault-Tolerant Design

With the volatile and dynamic nature of the IoT environment, the fault-tolerant design comes up to be essential to thwart the system from breaking up, particularly in such distributed ecosystem. Consider the scenario of many IoT devices collaborates on some tasks, if a device failure takes place in a particular device, then, other devices would typically be influenced. Federated learning researches emphasis mainly the privacy protection of different participants. Likewise, the acceptability of participants' devices has to be considered in a federated learning scenario. in another way, a control algorithm can be employed to decide the superlative tradeoff between global updates and local updates to acclimate to the drawback of local resources.

Some other works do not take into account the contribution of participants immediately, which did not impinge on the performance of FL. Another opportunity for enduring local device failures is a present reputation to enable reliable staff assortment where blockchain can be employed as a reputation management system, which can successfully thwart the malevolent attack and interfering [6, 10].

9.2.4.4 Learning Heterogeneousness

Typically, IoT data samples can be considered as the foundation of any efficient data-driven learning solution. accumulating erratically dispersed data across geographically distributed multi-party devices for developing a federated learning solution, will critically impact the final model performance [16]. Rational handling of data from various participants has a crucial influence on FL. To tackle the issue of numerical data heterogeneousness, the federated learning network can be primarily partitioned into three modeling approaches (1) particular device hosts its specific deep or shallow network; (2) federatedly learn a final solution appropriate for all participants; (3) trains pertinent shallow/deep networks for underlying chores [6].

9.3 Summary and Learnt Lessons

This chapter provides an in-depth discussion about the role of the federated learning for privacy preserved learning in IoT-based system, also, give valuable insights regarding the potentials of federated learning as follow:

- The chapter taxonomizes the federated learning solutions according to the sharing of feature and/or participant space, into three main categories i.e., HFL, VFL, and FTL. Also, it insightfully presents the merits and demerits of each category for IoT applications.
- The chapter argues the different techniques (secure aggregation, HE, and DP) for protecting the privacy of local parameters through the transmission between aggregation server and local participants. it details and subcategories of each one of these techniques are extensively debated.
- The chapter also explores the heterogeneity problem as a major obstacle limiting the applicability of federated learning solutions in real-world IoT environments. It give an insightful discussion about the main ways for addressing this problem along with their current progress and limitations.

References

1. S.K. Lo, Q. Lu, C. Wang, H.Y. Paik, L. Zhu, A systematic literature review on federated machine learning: from a sofware engineering perspective. ACM Comput. Surv. (2021). https://doi.org/10.1145/3450288
2. J. Verbraeken, M. Wolting, J. Katzy, J. Kloppenburg, T. Verbelen, J.S. Rellermeyer, A Survey on Distributed Machine Learning. ACM Comput. Surv. (2020). https://doi.org/10.1145/3377454
3. C. Zhang, Y. Xie, H. Bai, B. Yu, W. Li, Y. Gao, A survey on federated learning. Knowledge-Based Syst. (2021). https://doi.org/10.1016/j.knosys.2021.106775
4. A. Makkar, U. Ghosh, D.B. Rawat, J. Abawajy, FedLearnSP: preserving privacy and security using federated learning and edge computing. IEEE Consum. Electron. Mag. (2021). https://doi.org/10.1109/MCE.2020.3048926
5. M. Aledhari, R. Razzak, R.M. Parizi, F. Saeed, Federated learning: a survey on enabling technologies, protocols, and applications. IEEE Access (2020). https://doi.org/10.1109/ACCESS.2020.3013541
6. S. Abdulrahman, H. Tout, H. Ould-Slimane, A. Mourad, C. Talhi, M. Guizani, A survey on federated learning: the journey from centralized to distributed on-site learning and beyond. IEEE Internet Things J. (2021). https://doi.org/10.1109/JIOT.2020.3030072
7. P. Bellavista, L. Foschini, A. Mora, Decentralised learning in federated deployment environments: a system-level survey. ACM Comput. Surv. (2021)
8. A. Boulemtafes, A. Derhab, Y. Challal, A review of privacy-preserving techniques for deep learning. Neurocomputing (2020). https://doi.org/10.1016/j.neucom.2019.11.041
9. N. Waheed, X. He, M. Ikram, M. Usman, S.S. Hashmi, M. Usman, Security and privacy in IoT using machine learning and blockchain: threats and countermeasures. ACM Comput. Surv. (2021). https://doi.org/10.1145/3417987
10. W.Y.B. Lim et al., Federated learning in mobile edge networks: a comprehensive survey. IEEE Commun. Surv. Tutorials (2020). https://doi.org/10.1109/COMST.2020.2986024
11. K. Wei et al., Federated learning with differential privacy: algorithms and performance analysis. IEEE Trans. Inf. Forensics Secur. (2020). https://doi.org/10.1109/TIFS.2020.2988575
12. N.J. Gati, L.T. Yang, J. Feng, X. Nie, Z. Ren, S.K. Tarus, Differentially private data fusion and deep learning framework for cyber–physical–social systems: state-of-the-art and perspectives. Inf. Fusion (2021). https://doi.org/10.1016/j.inffus.2021.04.017
13. F. Sattler, S. Wiedemann, K.R. Muller, W. Samek, Robust and communication-efficient federated learning from Non-i.i.d. data. IEEE Trans. Neural Netw. Learn. Syst. (2020). https://doi.org/10.1109/TNNLS.2019.2944481
14. L.U. Khan, W. Saad, Z. Han, E. Hossain, C.S. Hong, Federated learning for internet of things: recent advances, taxonomy, and open challenges. IEEE Commun. Surv. Tutorials (2021). https://doi.org/10.1109/comst.2021.3090430
15. D.C. Nguyen, M. Ding, P.N. Pathirana, A. Seneviratne, J. Li, H.V. Poor, Federated learning for internet of things: a comprehensive survey. IEEE Commun. Surv. Tutorials (2021). https://doi.org/10.1109/COMST.2021.3075439
16. B. Gu, A. Xu, Z. Huo, C. Deng, H. Huang, Privacy-preserving asynchronous vertical federated learning algorithms for multiparty collaborative learning. IEEE Trans. Neural Netw. Learn. Syst. (2021). https://doi.org/10.1109/tnnls.2021.3072238

Chapter 10
Challenges, Opportunities, and Future Prospects

The central intention of this chapter is to discuss the primary security challenges in internet of things (IoT) environments with the main emphasis on the opportunities for deep learning for securing and maintaining the privacy of IoT-based systems. Bearing in mind the advanced computing paradigms. Then, the chapter discusses the security challenges meeting the cloud-based IoT, Fog-based IoT, and Edge computing-based IoT, while arguing the opportunities and future prospects for designing an intelligent security solution based on this paradigm. Follow, the challenges facing the design of deep learning-based security solutions are addressed from different perspectives along with potential trends and future directions that require more investigation. In a similar way, the opportunities of deep reinforcement learning are debated, and related future prospects are carefully charted. Finally, federated learning is extensively analyzed for limitations and downsides and possible remedies were suggested to be considered in the future.

This chapter chart the challenges, opportunities, and future prospects of securing IoT environments using deep through the following sections:

- Internet of things security
- Cloud-Centered Security Solutions
- Fog-Centered Security Solutions
- Edge-Centered Security Solutions
- Deep Learning for IoT security
- Privacy-Preserving Federated Learning

For convenience, the organization of the sections in this chapter follows the appearance of their topics in previous chapters.

Electronic Supplementary Material The online version of this chapter (https://doi.org/10.1007/978-3-030-89025-4_10) contains supplementary material, which is available to authorized users.

10.1   Internet of Things Security

Security Standardization: Thanks to the absence of standardization for IoT components (physical or digital) lead to an excessive amount of heterogeneous IoT products, service, applications, thereby suffering from security, compatibility, and interoperability issues. It worth motioned that the majority of IoT products are being constructed with no fundamental security standards [1]. Bearing in mind, the current IoT threats, there is a growing need to integrated security mechanisms into different categories of IoT devices. These mechanisms should principally be designed to satisfy the IoT security requirements. Nevertheless, considering the limited resources of almost all IoT devices necessitate this mechanism to be lightweight with low latency. Moreover, application-related characteristics, such as low production expense and low power utilization are deemed to be the restricting considerations in creating a standardized or generalized security solution for all the IoT products. Therefore, it is highly required to have worldwide IoT standards imposing the smart things automatically and easily fulfill the stated security requirements [1].

Software or Code Integrity: Several methods have been designed to confirm the integrity of IoT devices. Nevertheless, the highly trustworthy solutions are hardware-founded requiring the implementation of extensive verification procedures in a reliable IoT environment. Bearing in mind the extent of deploying the cheap and tiny IoT devices, the production of protected hardware based IoT objects to be used in crucial infrastructure is impractical at all. Therefore, it is highly necessary to investigate a protected software-based approach that could be simply installed on resource-limited IoT devices with the plasticity of timely updates. One More conceivable challenge is that the new generation of IoT (5G, 6G) would contain a massive number of heterogeneous IoT entities. Thus, the efficient detection and rectifying any malicious code or software adjustment will be very challenging in wide-ranging dynamic and heterogeneous networks [2]. A more intelligent swarm attestation mechanism might provide a solution for this challenge.

10.2   Cloud Computing Based Security Solutions

DDoS attacks in cloud-backed IoT: Cloud computing delivers the virtual resources (i.e., storage, computing, etc.) to the planned IoT users, upon request. The scale of standard cloud applications has been widely expanded with the integration of cloud and IoT capabilities [3]. The cloud- backed IoT adds a set of new features (i.e., smart things, advanced communication, etc.) to the standard cloud system and delivers innovative examining and robust processing abilities, and many other benefits. In addition to the gained benefits, this integration comes up with countless security problems such as reliability, heterogeneousness, privacy, etc. For instance, with the advent Cloud-backed IoT, the potential of DDoS attacks is expanded vastly, while the detection of these type of attacks remain in its early phases. Thanks to the recently incorporated facet, the present DDoS prevention or detection methodologies might not be appropriate for the cloud- backed IoT environment. Therefore, additional research and investigation in deep learning solutions can pave the way toward more intelligent solutions for mitigating the risks of DDoS attacks in cloud- backed IoT environments [4].

DDoS and communication overhead: According to current literature, it worth noting that multiple defense attempts have been proposed for different categories of DDoS attacks (i.e., low-rate, high-rate). Most if not all of them necessitate collaboration of the middle routers, which in turn cause an additional operating cost on the middle nodes and can negatively impact the performance of the methods as well. Triggering any DDoS attack cloud-backed IoT environment mostly involves concealing the actual attacker's location by applying IP spoofing practice. Hence, the recognition of spoofed DDoS attacks is a critical but challenging task. The inclusion of intermediate routers in the present efforts for detecting spoofed DDoS attack usually result in high communication overhead limiting the ability to detect various DDoS attacks in an efficient way. Even Though some attempts have been made to lessen the overhead in cloud communication [5], however acceptable results still unreachable. Thus, more efforts should be devoted optimize the communication overhead.

DDoS and Computational complexity: Detecting low-rate DDoS attacks often entail methods dependent on Discrete Fourier Transform (DFT) for energy spectral estimate. In which the calculation of N-point DFT involves $N(N-1)$ addition operations and N^2 multiplication operations. Hence, the computational complexity has to be $O(N^2)$. This typically influences the performance of the applied method making it unacceptable for time-sensitive cloud applications or services. Thanks to the commitment of intricate computations, the DFT-based methods demonstrated laborious solutions and gives an inadequate performance as the DFT is inefficient for the big values of N. Therefore, a more academic or industrial effort has to be devoted to lessening the complexity of DFT computation or to develop a new computationally efficient transformation procedure [4, 6].

Interior cloud DDoS attack: The majority of solutions for detecting DDoS in cloud environments assume the IoT traffic arriving from the exterior network is the key target for assessment. The exterior network describes the network beyond the boundaries of the cloud infrastructure. Hence, applying these solutions enable capturing exterior DDoS attacks only. In cloud based IoT, DDoS attacks can be initiated with virtual machines in the same infrastructure. This case is identified as internal DDoS attacks which still unhandled or slightly explored in the previous works. Therefore, the probability of DDoS attacks remains viable despite the currently applied solutions. The scrutinization of interior and exterior traffic of the cloud network without enforcing extra communication burden is crucial and challenging. As a promising solution, an improved deep learning model can be trained to discriminate and detect both kinds of cloud DDoS attacks [4, 6, 7].

Instantaneous cumulative analysis of traffic flows: The cumulative stream of traffic describes the data flow received from the exterior and interior, valid and malicious entities. It worth noting that the research literature did not address the instantaneous cumulative traffic, as the majority of presented solutions is experimented only within the simulated environment completely differing from the genuine cloud environments. Hence, the appropriateness of these solutions to an extensive-scale cloud network is difficult if not impossible to be convinced. In the case of cloud intrusion detection systems, analysis of instantaneous cumulative traffic is critical for decreasing the probability of IoT attacks [8, 9]. To enhance the performance of attack detectors, some resource efforts have been addressing the detection in real cloud settings, and the initial findings show that entire network traffic is required for analysis. Therefore, a considerable investigation should be performed in the future to solve this challenge.

Commercial loss: These Days, attackers are entranced in the direction of the low-rate DDoS attacks by constantly attempting to uncover new crafty attack initiating tactics. Hence, the recognition of this category of DDoS attacks is arduous. For example, the Economic Denial of Sustainability (EDoS) attack seeks to make use of the automated scaling nature of the cloud computing and triggers needless deployment of the novel virtual machine cases. Failing to identify Such kinds of DDoS attacks within a time threshold, may possibly trigger an enormous commercial loss. Current detection or countermeasure solutions for the EDoS attacks emphasize alleviating the attack in a reactive manner. Reactive solutions alleviate the attack following its successful execution [10, 11]. Hence, these solutions might expand the charge of attack alleviation and subsequently disturb the commercial factors of cyber-physical systems. Therefore, preemptive deep learning solutions have to be developed to increase the possibility of early detection of these sorts of attacks.

SDN-based cloud solutions: Cloud-backed IoT environments turn out to be a major ground for different IoT attacks especially DDoS attacks because of the different set of relevant features, which necessitate robust security methods counter to DDoS attacks. A promising direction for this may include the exploitation of Software Defined Networking (SDN) technologies. With the latest innovations in SDN, this emerging model has earned huge interest in the domain of IoT networking. SDN uncouples the data surface and management, rationally consolidates the networking information and summarizes the IoT network's intricacy for different applications [5, 10, 12]. The integration of this hopeful technology may offer modern enchanting opportunities for the detection and defense against DDoS attacks in cloud-backed IoT ecosystems. SDN-founded cloud computing is an emergent paradigm to enable SDN controller to take the control of the infrastructure of IoT network in such a way that promote delivering the networking as an on-premises service. The characteristics of SDN like unified control, computerized traffic analysis, active modernizes in dispatching regulations, etc. render it simpler to protect system from DDoS attacks. In Spite Of many advantages, the SDN turned out to be susceptible to DDoS attacks which ought to be tackled during the development of security solutions. Some research efforts have been exploring SDN-based solutions for different IoT attacks including DDoS attacks, however, these solutions are stagnant and incapable to adapt their tasks to match the type or gravity of DDoS attacks. Thus, they are deemed to have an inadequate achievement rate. Therefore, new dynamic SDN-based solutions should to be developed address the aforementioned challenges in Cloud- assisted IoT environments.

DDoS-for-Hire service: service providers can be viewed as the new performers in the ground of DDoS attacks taking the responsibility of delivering DDoS-for-hire [13] service upon fee foundation. The attacker can abuse upon-request self-service and speedy flexibility aspects of cloud-backed IoT environment to develop a formidable botnet and delivers DDoS-for-hire as a service. This cloud paradigm is executed through the usage of huge volume of cloud storage and computations. Since this cloud is under the control of attackers, it is commonly identified as black-box clouds. Thanks to the emergence of the DDoS-for-hire service, a great increase might be witnessed in the proportion of DDoS incidents in a cloud-backed IoT ecosystem. It might come to be simpler to initiate DDoS attacks even for beginners. Therefore, the development of security solutions (i.e., deep learning-based) solutions to detect and prevent this kind of attack is of great importance either in academia or industry [14]. The alleviation of DDoS attacks applied utilizing DDoS-for-hire service is costly since extra resources are necessary to thwart the attacks. Therefore, developing a resource-efficient solution for this form of attack is difficult and should be considered in future prospects.

Detection Near Impossible (DeNy) attack: A modern but complex attack developing as a public research challenge threatening the security of cloud systems. The DeNy attack is craftier and utilizes fluctuating percentage of attack traffic. Hence, detecting it with the prevailing security solution is almost impractical. This attack seeks to transmit only normal IoT traffic from a huge number of sources varying in location. This kind of attack is untraceable because of two characteristics namely fake alarms, and benignness. Benignness indicates that the traffic of the attack has no abnormalities. Hence, the attack traffic constantly produces false positives. More, the traffic representation of DeNy attacks varies from those of the low-rate DDoS. So, it might stay unnoticed by the detection solution employed for low-rate DDoS attacks. Therefore, the development of a deep learning security solution for the DeNy attacks can be viewed as a challenging and interesting task that necessitates improving the current avoidance, recognition, and alleviation techniques [4, 13].

Cloud-backed IoT forensics: from the IoT forensics perspective, six main challenges have to be considered for designing a robust investigation framework which is indicated by the mind map presented in Fig. 10.1 [15, 16]. This includes evidence identification, evidence collection, evidence preservation, evidence analysis and correlation, attack attribution, and evidence presentation. With the advent of cloud-backed IoT systems, these forensics aspects become more complex and challenging to standardize. Also, it offers a great opportunity for cloud providers and researchers to design and establish new business models, hence, paving the way toward the development of the IoT Forensics-as-a-Service (IFaaS) paradigm as an improvement over Digital Forensics-as-a-Service (DFaaS). Getting the entire IoT ecosystem secured, however, cannot considered a simple mission. Different from traditional

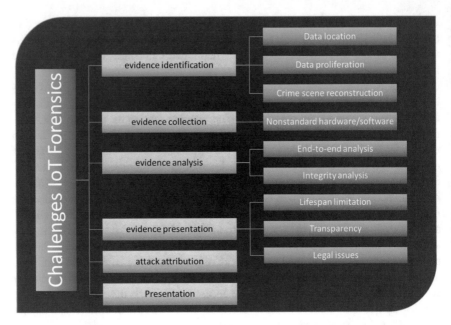

Fig. 10.1 An overview of challenges of the IoT Forensics

computers, which depend on conventional network security standards (i.e., firewalls), the communication in cloud-backed IoT involves large number of protocols, IoT devices, facilities and specifications.

10.3   Fog Computing Based Security Solutions

Fog computing cannot be considered a solution capable of addressing the security gaps of cloud- backed IoT without incurring any problems. The paradigm is still in initial phases, and thereby the large-scale deployment remains enormously constrained. This further reveals multiple open challenges related to resource offloading, privacy, functional federation scheme, and other security challenges [17]. While it is believable that the different deep learning solutions can provide a remedy for these challenges by fulfilling the main security requirements, it worth noting that different learning solutions bring many limitations and tradeoffs, and systematic investigation of the reimbursements and downsides of Fog computing could be performed after abundant Fog-enabled IoT systems become available to explore [18]. Nonetheless, an ideal analysis could be obtained according to the recent cutting-edge solutions. Consequently, this section briefly argues the most possible challenges facing the development and the deployment of deep learning security solutions on a Fog-enabled IoT environment.

First, Fog-enabled IoT presents an additional load to the maintenance workforces, and will probably necessitate specific training, rendering it more expensive compared with the cloud- backed settings. While Cloud- backed IoT usually maintained using a team of specialists at the side of the cloud provider, the introduction of Fog-enabled IoT transfers this accountability to the system's end-users. The proliferate of functionality across the Cloud-to-Thing's architecture obscures this even more. When a security problem has taken place with a well-established section of Fog infrastructural software, the maintainers are accountable for the overall range to upgrade their software, compared to necessity for upgrading the cloud infrastructure, under the management and maintenance of a single entity. Unfortunately, the failure of a maintainer of a particular Fog node unavoidably endangers overall IoT entities using that node [19].

Besides, performing and identity authentication whilst guaranteeing low latency of IoT services can be deemed as a significant challenge of fog computing. This can be more challenging especially in the case of real-time applications, mobile IoT devices, a decentralized configuration of fog nodes. Despite the availability of a wide variety of identities authentication policies, however, they fail to consider the mobility of the IoT devices. The possible remedy for this is the application of blockchain to control access in fog computing. In this solution, the fog nodes could the complete connections for the Blockchain and could efficiently sustain the security of verification process and approval evidence utilizing consortium signatures or property-centered encryption [20]. However, the high latency of blockchain solutions still a limiting factor for solving this challenge.

In contrast, the consistency of the access control strategy is challenging in the case of manifold customers utilize the IoT entities to gain access to immediate public facilities. The strategy might entail authentication of IoT devices, client management procedures, and key administration techniques for the fog severs. Though protection is an indispensable requirement of IoT environment, nevertheless, the restricted computing, storage, and energy resources of IoT devices limit the applicability of traditional cryptographic approaches. Hence, it is highly needed to develop a lightweight security approach that can fulfill the needs of real-time services for fog-enabled IoT applications [21].

Additionally, the disseminated nature of Fog-enabled IoT is likely to obstruct incident response. This is believed to express itself in manifold ways: required security experience could not be accessible on-site, thereby dedicated incident response crews should be invited from outer coalitions. Moreover, complicated threats/events could necessitate collaboration among various IoT units with the main aim to realize insightful forensic analysis, which may not continually be conceivable or include much overhead.

Furthermore, affinity among fog devices may possibly lead to substantial problems. With the absence of well-established standards or weak follow of standards, it will be extremely difficult to encounter the strict requirements imposed by cyber-physical IoT environments with heterogeneous nodes from a variety of providers that are unable to interoperate professionally and truthfully. Which in turn, can adversely influence the capability to amalgamate and offload deep learning responsibilities to other IoT entities across the local network, and possibly contravene security strategies when some IoT nodes become incapable to fulfill the essential obligations.

Finally, with the advent of the Fog-backed IoT environment, new level computation was introduced by bringing the resource closer to the IoT devices, which scale up the IoT environment to deliver services to a wider range of devices compared to the cloud-backed configurations. The expansion typically increases the complexity of previously mentioned challenges of IoT forensics presented in Fig. 10.1 [15, 16]. Moreover, the lack of standardization and maintenance specialists makes the development of fog computing forensics frameworks a promising future direction.

10.4 Edge Computing Based Security Solutions

Context-Informed Security: Context perception of mobility IoT technologies take the advantages of recent improvement in sensing capabilities of smart devices to empower ambient intelligence such as Microsoft Cortana, Google Now, and Siri [22]. The most conventional exploitation of circumstantial information is the establishment of individualized and/or location services using the captured geographic coordinates. These cooperative context-informed applications, often elevate many concerns related to the privacy and security of IoT data, especially in edge computing (EC) environments. Nonetheless, exploiting the context information about the IoT devices (i.e., mobile, or stationary) and protocols to gather security-associated information facilitates the deployment of adjustive security solutions to transmission protocols [23]. This mandate distributes security solutions to be deployed at the edge nodes in an EC- backed environment, explicitly to deploy an independent security function in MEHs for detecting abnormal system behaviors based on the security considerations originated from context information. The Security as a Service proposals suggested for EC- backed IoT environments are systematizing the installability of security solutions, as a self-sufficient and smart task to improve the context-awareness [24]. Therefore, developing an intelligent context-informed security solution can pave the way toward addressing much of the security issues EC- backed IoT environment [25].

Microservices-based design: Microservices denote a structural system design that designates an IoT application or service as a compilation of complementary modules services known as microservices that present lightly combined, extremely maintainable, and autonomously deployable characteristics [26]. To assess the application compatibility with the microservice-based design, a pattern language can be employed. Nevertheless, the microservice design does not streamline the system operations. It decomposes the logic of application into manifold reduced modules hosted on a container, leading to a much more complicated network communication model between microservices [27]. In the perspective of mobile edge computing, the IoT device Apps and mobile edge Apps might be viewed as microservices associated to one another for recognizing a variety of practices implicated in the MEC operation, security can be provided as a service. Microservice-founded authentication methods are crucial for guaranteeing the privacy of applications and the reliability of EC connectivity procedures. In this context, the osmotic computing evolving paradigm seeking to accomplish a harmonious relocation of edge and cloud computing platforms [28, 29]. Motivated by the Osmosis sensations, the most remarkable about this paradigm is the dynamic orchestration of services and microservices throughout cloud servers and edge nodes. Microservice implementation and resettlement activities can be protected efficiently from an osmotic computing structure that implants consistent security strategies shared among the edge and cloud servers. Few studies tried to investigate the microservice-based deep learning solutions, however, this topic still in its initial phases and can be viewed as a promising direction for efficient deployment of powerful deep learning security solutions in an EC- backed IoT environment.

Mechanisms Orchestration and Standardization: thanks to the heterogeneousness of both IoT software and IoT hardware in an EC-backed IoT environment, it turns out to be important to competently orchestrate diversity of privacy and security solutions. This can be accomplished by creating adaptable and cohesive security systems, principles, policies, and strategies that empower heterogeneity, interoperability, compatibility, and demonstrate resistance against most IoT attacks [30]. Developers and service providers have to design such a cohesive security framework considering the delicate functioning terms and variations of the fundamental IoT devices and edge odes, as such aspects significantly affect the implementation and execution of EC-backed IoT framework. Moreover, considering that there are a variety of third-party colleagues engaged in designing EC-backed IoT networks, i.e., network suppliers, product developers, and service suppliers, the challenge of developing cohesive security and privacy framework has come to be even more difficult [31]. These partners ought to collaborate to develop interoperable security solutions for simplifying the streaming of IoT information with a high level of security [32]. Therefore, privacy and security policies are essential in supporting the design of a secure EC-backed IoT environment.

10.5 Deep Learning for IoT Security

Insights into deep learning: Deep learning solutions offered a great evolution of the methodology through which a computing device solves a particular challenge by training the machine to make an intelligent decision based on previous experiences. However, in spite of the advancement achieved by deep learning models in different disciplines, a philosophy that might explain why and how deep learning models operate based on their building was not investigated till now. This philosophy may be important in grasping the amount of information or the total layers necessary to reach the anticipated performance. The philosophy could also enable optimizing the resources consumptions of deep learning models i.e., time, space, power [33], in that way delivering powerful but lightweight deep learning solutions that are helpful for resource-restricted IoT devices. The development of a lightweight deep learning algorithm is a substantial step in the direction of the design of on-board security frameworks for different IoT entities. Therefore, this topic is the most interesting challenge that should be addressed in future plans.

The dataset scarcity: The common objective of deep learning is catching the hidden representations from the existing incomplete training data and then building a model to categorize the incoming IoT traffic samples based on the learned representations. In view of this, an important matter to explore is whether the quantity of observations necessary to train the deep learning model adequate for the generalization of these models on a new input in a specific IoT territory [34]. From the standpoints of IoT security, the main challenge confronted by deep learning generally, and the supervised models specifically, is the way to obtain or produce practical and high-value training data that encompasses a rich set of potential attack categories. A high-value training data can be considered as a crucial factor to efficiently train the deep learning-based security solution. The more the training data is rich, inclusive, and varied, the more the performance is well [35]. It has to include a set of features and classes that represent the majority of the plans of real-world IoT attacks for the reason that this training data form the foundation for acquiring model knowledge. This case often has an explicit impact on the model performance. Provided the dynamic nature of IoT systems generating huge quantities of traffic data, the data superiority maintenance, and real-time data flooding persists as challenging tasks.

A critical future prospect is the development of crowd-sourcing approaches to generate rich datasets that contain almost all attack forms for the effective training of deep learning models. Moreover, such datasets could be employed to benchmark the performance of recently suggested deep learning solutions for recognizing and discriminating different IoT attacks. Despite the importance of creating cooperative IoT attack datasets that can be constantly upgraded with the lately discovered attacks, it is technologically challenging owing to the large multiplicity and heterogeneousness of IoT devices. Furthermore, privacy concerns predominate as the datasets might include confidential or important information that is not intended to be disclosed to the public community, particularly for manufacturing, military, and medical IoT devices.

IoT Security Solution based on low-quality data: The majority of presented deep learning-based security solutions are generally experimented and evaluated using high-quality IoT data i.e., cleaned, preprocessed, and formatted. Nevertheless, cyber-physical IoT environments encompass heterogeneous internet-connected devices, and broad-range streaming, thereby entailing the likelihood of high-level noise and distorted data to be presented in such environments [36, 37]. Thus, learning to secure IoT based-systems necessitates efficient deep learning solutions that can effectively cope with and learn from limited-quality data, especially in situations where superior-quality training dataset is essentially impracticable. Consequently, multi-modal, and efficient deep learning solutions need to be developed to protect IoT-based systems having huge-range streaming, diverse and noisy IoT samples.

IoT security data Augmentation: Instinctively, the wealthier the data that deep learning models trained with, the more correct they could be [38]. While acquiring a larger size dataset is comparatively straightforward in specific disciplines i.e., natural language processing and computer vision, however, obtaining a huge dataset for deep learning is comparatively challenging in the realm of cyber-physical security of IoT environments. Thus, investigating alternate methods for acquiring considerable volumes of rich and representative IoT samples become of great interest. The augmentation task is employed to enlarge volume of small dataset by synthesizing a new data instance similar to the original data samples. In this context, the small volume of accessible IoT traffic samples may be exploited to synthesize and engender larger set of training instances. The major issue with augmenting IoT data is generating reliable set of traffic instance following suitable distribution of data across different attack categories, which is typically requiring domain experience [39, 40]. In view of this problem, appropriate augmentation techniques should be investigated for enhancing the learning capabilities of the deep learning model and the performance of dependent security solutions.

Zero-day attacks: The key advantage of deep learning solutions over conventional security solutions (i.e., signature-founded solutions) is their ability to distinguish zero-day attacks. Zero-day attacks refer to a kind of lately developing attacks that were is formerly unknown to detection mechanisms. They are known to have fluctuating capacities, similar to metamorphous malware which routinely renovate itself every moment they are disseminated or broadcasted. Thus, identifying these kinds of attacks using the conventional solution is tough or even impossible [41]. The number of evolving IoT attacks like zero-day attacks is constantly expanding at a frightening ratio. For instance, different variants of Mirai botnet are come to be a disturbing danger to the security of IoT environments [42]. The introduction of the latest variants of Mirai botnet, Satori, validates that other malevolent IoT botnets are evolving to exploit established and zero-day susceptibilities [43].

Besides, latest beginnings of Mirai indicate that IoT malware remains to evolve owing to the open-source Mirai's code, which tolerates the developers or attackers to generate different deviations of Mirai that make use of established and zero-day susceptibilities to attack IoT devices [44–49]. In addition, smart watching and management of IoT security deliver an imperative solution to the newly generated attacks or the variations of a zero-day attack. Deep learning is a promising analysis means for learning regular or irregular behavior based on collaborations among the devices and systems within an IoT environment. Input samples from each component of an IoT system and smart things devices could be accumulated and analyzed to verify regular patterns of interaction and therefore recognize malevolent behavior at an initial phase. Furthermore, in view of the ability of deep learning solutions to learn from occurring IoT samples to smartly foresee imminent unfamiliar observations, these solutions come up with the capability to forecast new IoT attacks, which almost are straightforward transformations and mutations of earlier kinds of IoT attacks. Consequently, IoT security systems need to advance from the simple enablement of protected interaction among devices to knowledgeable deep learning-based security solutions.

Enduring Learning: Dynamism is treated as one of the major attributes of the IoT environment in which numerous modern objects enter and many entities depart the system provided the plentiful and heterogeneous IoT devices employed to handle various applications and circumstances. Accorded that the nature of IoT, standard constructions and patterns of IoT ecosystems might significantly adjust over time, and potential IoT attacks affecting the IoT-based environment might similarly continually fluctuate with time. Hence, differentiating standard behavior from anomalous behavior may not be forever pre-specified. Therefore, the regular renovating of security solutions is necessary to cope with and realize IoT updates [50]. In a typical IoT ecosystem, the short-time learning of IoT attacks from the received samples might be ineffectual for long-time security. Therefore, the idea of enduring learning could handle the practical implications of long-standing real-world IoT services or applications. The notion of enduring deep learning is pointed on the way to the building of a model that can continually be retrained for learning new emergent features associated with different attacking behaviors [51]. In other words, the model ought to be capable to continually acclimate to adjustments in the IoT environments [18, 51]. Recent offers in this direction show that with additional training, the learning model experience the recent features of notorious DDoS attacks, which further improve the chances of discovering established and anonymous DDoS attacks. Deep learning learns hidden representation from the training IoT samples and subsequently identifies the features of zero-day attack that are similar to the characteristics experienced during the training. Consequently, regularly updating training instances is essential for cultivating efficient security solutions for real-world IoT applications.

This includes the development of an intrusion detection system that can automatically update the directory of the discovered IoT attacks as soon as a novel attack appears. To this end, IoT-based intrusion detection has to be experienced with robust deep learning strategies that enable updating the training parameters of the model in real-time in such a way to keep the model's knowledge up to date, in incremental deep learning can be viewed as a promising solution to this opportunity. Incremental learning refers to the strategy for retraining the deep learning architecture using both formerly seen and unseen data with the main aim to expand the knowledge of the current model. In other words, it seeks to assure the permanency of the learning procedure via routine model update depending on a new obtainable collection of data [52, 53].

Transfer learning: It describes the concept of transporting expertise from a territory with adequate rich dataset to a realm with an unsatisfactory amount of data. The central objective of transfer learning revolves around reducing the workload and time needed for the new learning operation [54]. The major interest in transfer learning consider the amount of domain expertise that could be transported in form of common knowledge between different data domains [55]. Consequently, it is effective to exploit this transferred knowledge during the design of a new security solution. Meanwhile, the knowledge transfer should be avoided under the situation in which it does not retain any significance to the underlying domain.

Integrating the conception of transfer learning into an IoT environment consisting of various components like IoT devices, sensor networks, fog computing, edge computing, and cloud computing, could be beneficial for empowering the security of IoT-based systems [56]. The protection of those components has already been comprehensively analyzed by generating properly-formed training instances of various attack classes [57]. Therefore, when achieving transfer learning effectively from IoT entities, so such learning might significantly enhance the performance of the security solution without incurring extra efforts and charges in assembling training datasets.

Interdependent, interrelated, and collaborative ecosystems: The following discussion argues the opportunities for using deep learning solutions to alleviate domestic security concerns evolving by the configurations of IoT ecosystems as they are inter-reliant, interrelated, and collaborating environments.

Again, with the speedy growth in the scale of IoT production, the cooperation among IoT entities is coming to be more independent, in such a way that involves lowered human engagement. IoT entities no more merely cooperate with each other as with traditional standard networks. Lots Of contemporary IoT entities are intended to realize the dream of a smart city, where multiple devices are automatically regulated by other devices or hinge on the functioning states of other entities in the adjoining environment. The gain of applying deep learning models in protecting IoT entities in smart environments is that they could be established to go beyond merely comprehending the functioning behavior of particular IoT entities to comprehending the functioning behavior of whole systems and the related devices [33].

Additionally, the IoT environment comprises billions of internet-connected devices; hence, not just the attack surface, yet the extent of the incident ought to be taken into account in IoT ecosystems. Given such intensively interrelated IoT devices, a contaminated entity can yield a devastating attack that overwhelms a huge number of IoT things to a significant extent, even influencing a large portion of a smart city.

In view of interrelated nature of IoT, the advantage of applying deep learning models for securing IoT entities is that they could deliver intelligence to systems for discovering anomaly behaviors of IoT entities individually or collectively and hence afford automated defense at an early phase of the attack. This policy might alleviate the effect of the attack and thereby enable training for the inhibition of forthcoming incidences of comparable attacks based on a consistent interpretation of the present reasons.

In the same time, deep learning provides an effective solution for securing collaborative IoT such as social IoT, which always prone to more complex and multi-facet IoT attacks. Appropriate directives ought to be launched for things to decide on their suitable friends for the reason that these directives influence the quality of services developed and provided over of social networks [58]. The evolution in social IoT raises many crucial security issues and privacy matters concerning the leakage of confidential information of involved entities. Therefore, deep learning is likely to donate to ensure the security of integration of social IoT environments. Nevertheless, this direction is still in its initial stages and requires additional investigation.in the future.

Deep Learning misuse: Recent advances in deep learning approaches have facilitated them to be employed for breaching cryptographic applications. Some studies show that machine learning algorithms can effectively break cryptographic systems better than template attacks. In later times, deep learning has been demonstrated more efficiently than machine learning for breaking cryptographic systems. Convolution networks and autoencoders are typical examples that have been experimented for these purposes. On the other hand, some studies demonstrated that recurrent networks can be employed to learn to decrypt the ciphertext. Unambiguously, an RNN based on LSTM with 3000 units could be trained to capture Enigma decryption role thru learning active interior representations of these ciphers; the findings advocated that recurrent networks could use the polyalphabetic ciphers to extract and learn the algorithmic patterns intended for cryptanalysis [59]. Therefore, future prospects should take into account these capabilities of deep learning as a main aspect while developing IoT security solutions.

Privacy of deep learning: Though the great success achieved by different categories of deep learning solution i.e., supervise, unsupervised, and semi-supervised. However, they prone to leading attacks and unable to protect the privacy of underlying training data [60]. On the other hand, the federated, distributed or even decentralized deep learning solutions tried to address these issues, however, they still are simply damaged and incapable to retain the training data privately.

Some efforts have made to develop an attack to comply with the instantaneous landscape of a training operation, where an adversary was tolerable to train a generative adversarial network that generates fake observations analogous to those in the underlying training data, which were presumed to be private; the fake generated observations were presumed to follow the same distribution of the original private dataset [61]. Hence, the models are susceptible to possible attacks during the generation phase. Therefore, attackers could develop a deep learning framework that can recognize the way the deep learning detection solutions operate and then create attacks that might not be effortlessly detected. Despite different late attempts to address this challenge, however, from the research perspective, these points still in their early stages and requiring more effort and investigations in the future.

Security of Deep Learning: Multiple works have lately investigated a variety of attacks that may be initiated against deep learning solutions. The deep learning models are vulnerable to various attacks that either declines the security performance or endanger confidential data. A typical example of these possible attacks involves poisoning, evasion, impersonation, and inversion attacks [58]. First, Poisoning attacks seek to insert malevolent examples with erroneous annotation into the training data with the main aim to adjust the distribution of training data, thereby diminishing the discrimination power of the classifying different categories of system behaviors, which subsequently decline the model performance. Such attacks can be possibly initiated against deep learning models that require enthusiastically modernize their training data and learning parameters to cope with features of new attacks [62]. Second, the evasion attack in the basis of generating adversarial observations by adapting the attack structures to be somewhat dissimilar from the malevolent observations employed to train deep learning model; therefore, the possibility of detecting the attack get reduced, and even the attack evades the discovery, thereby dipping the performance of the system curiously [63]. Third, the impersonation attack, which struggles to simulate the original data observations to betray the deep learning models to categorize the original observations with various incorrect labels from the impersonated ones [64]. Finally, the inversion attack seeks to take advantage of the application program interfaces given to the consumers of the deep learning framework with the main aim to roughly aggregate the needed information regarding the pre-trained models [65, 66]. Later, this mined information is employed to carry out reversal engineering to acquire the confidential data of customers. Such an attack contravenes the customers' privacy by investigating their data, which are confidential under some instances (e.g., patients' medical data), injected in the deep learning training stage [62, 65]. Therefore, the security of deep learning models that hold possible applications to protect IoT devices has to be realistically and satisfactorily protected against adversarial leakage of their gradient information.

Blockchain and Deep Learning: Blockchain is a recently evolved technology that employs cryptography to protect various network transactions. A blockchain delivers a decentralized ledger of operations, in which every network node is kept informed [67]. The network is designated as a series of IoT entities that all want to approve a transaction prior to it could be validated and documented [68]. Specifically, the blockchain can be basically defined as a data structure to enable the fabrication and dissemination of a 'tamper-proof digital ledger' for data interactions [69]. The decentralized design of a blockchain is adversative to intrinsic security challenges in a centralized design. Employing decentralized database design, the task of transaction authentication hangs on the authorization of several system parties instead of a single authority, which is a popular case in centralized platforms. Thus, blockchain platforms could render transactions fairly more reliable and translucent than those in centralized platforms. The distributed nature of IoT-based systems makes the blockchain a promising solution for securing IoT environments. However, the latency issue remains the main limiting factor for the applicability of blockchain in real-world time-sensitive applications. On the other hand, Deep Learning is interested in empowering the training devices to learn from real-life observations to later perform independently and cleverly. The objective of deep learning models is to transform traditional computing machines to be intelligent machines. The streamlined designations of both technologies disclose that a synergic relation may be achieved by integrating them to realize a completely operational IoT security system. First, deep learning may support blockchain technology to realize intelligent decision-making, enhanced assessment, purifying and understanding the IoT devices and data in a network to enable the efficient application of blockchain ledger for improved confidence and security of IoT services. Next, blockchain might help out deep learning by delivering a huge quantity of data because of the decentralized nature of blockchain that emphasizes the significance of distributing the data across multiple networked nodes. The introduction of big IoT data increases the need for developing a powerful deep learning solution. Thus, by the growth of the amount of information to be analyzed, especially security-associated IoT data, the truthfulness of blockchain-enabled deep learning models may be significantly improved and generalized to develop more reliable security solutions.

Computational complexity: Most IoT things having limited resources such as CPU, memory, storage, and energy, which are needed for training and deploying deep learning models. this limitation establishes a critical bottleneck in the acceptance of deep learning for real-time on-the-ship implementation. The existing techniques computation offloading and implementation in the cloud or fog environments experiencing elevated wireless energy burden. Likewise, the practical application of these kinds of techniques often depends on the network circumstances. Therefore, the cloud and fog resource offloading may be unachievable in case of weak network connectivity, causing the unapproachability of the applications. Another modern solution that may advance the implementation of deep learning for IoT security is the design GPU empowered edge computing and /or mobile GPU devices. Nevertheless, deploying GPUs on a mobile device still exhibits a high battery consumption rate [70, 71].

Moreover, boosting GPU-based deep learning approach and recommending an effective offloading scheme are essential in improving the implementation of deep learning solutions to improve the IoT security of IoT-based systems in cloud-backed, fog-backed, or edge-backed environments [70]. Furthermore, the design of supervised, unsupervised, or semi-supervised deep learning frameworks should impose substantial consideration for time and computational efficiency. Designing instantaneous discovery and security solutions is essential for fulfilling the security requirements, especially for dynamic, broad-scale, and heterogeneous IoT ecosystems. Therefore, decreasing computational complexity is the main factor for determining the real-world applicability of deep learning-based security solutions in the future IoT-based system.

IoT Security Trade-offs: The present security trade-offs, like that between privacy availability trade-off, security-safety trade-off, are an additional contest to the realization of a vigorous security solution for IoT ecosystems. Additionally, the significance of different security trade-offs varies according to the underlying IoT application. For instance, on the internet of medical things environment, systems have to offer a security solution, however, it must also support the elasticity of being easy to use under emergency circumstances. If a patient unexpectedly undergoes an emergency, effortless admission to the implanted IoMT apparatus is the key precedence for conserving the patient's life. Thus, it is highly recommended to design a framework that exhibit equilibrium between vigorous security solution to secure the implanted IoMT device and ensuring flexible access to the IoMT devices throughout an emergency. A suitable equilibrium between security and other application-related factors is an essential consideration to be regarded in the design of secure IoT applications [72]. Deep learning solution primarily seeks to deliver intelligence as well as circumstantial knowledge to IoT devices; hence, these solutions can be more effective for alleviating security trade-off concerns compared with conventional access control techniques. In the same way, different set of IoT applications exhibit a variety of trade-offs in accordance with diverse implemented environments. Provided the necessary security degree and trade-offs in particular IoT applications, the security strategy has to comply with various operation modes of the underlying IoT applications [58]. Future research may utilize the intelligence capacity of deep learning security solutions that would efficiently fulfill different security trade-offs in IoT environments.

10.6 Deep Reinforcement Learning

IoT environments are known to be dynamic, complex, and heterogeneous in nature that yields a massive number of transactions on daily basis. The millions of networks connected IoT devices have been rendering the development of security solution as a high-priority research challenge. Deep Reinforcement Learning (DRL) was demonstrated effective for securing the IoT environment from a variety of attacks. Nevertheless, some challenges and open issue remain unaddressed, which is going to be pointed as follow.

Resistance against Adversarial reinforcement learning: Considering the adversarial environment can be a major challenge and enthusiastic field of the research area in the context of reinforcement learning-based IoT security. Few efforts have been devoted to exploring the potential of reinforcement learning to detect or prevent adversarial attacks. The IoT environment may contain some adversary that constantly attempts to prevail over the agent striving to realize and understand the environment. In a multi-agent configuration, a single may take the role of adversary. Thus, a novel way is highly required to guarantee that the realized policy is vigorous against any ambiguous modifications in the environmental state. It is highly challenging to develop an intelligent security solution to enable the agents trained with reinforcement procedures while maintaining resistance against any adversarial attacks [73].

Deficient Perception Dilemma: In an IoT environment, the agents could not always have ideal and full awareness of the environmental state. This can be caused by two factors. First, restricted sensing facilities of sensors in the physical layer. Second, information loss owing to restricted or interrupted communication ability in the network layer of the IoT system. In view of this, a significant challenge in developing DRL solutions is to learn with partial perception or incompletely discernible states. The Markov Decision Process (MDP) model becomes no longer acceptable, as the state information is unsatisfactory to sustain the determination on optimum action. The action may be enhanced when extra information is accessible for the agent along with the state information. Despite the effort made in developing DRL-based security solutions in IoT environments, it is considered an infancy topic with many challenges. First, an agent in a partially observable Markov decision process (POMDP) demands to decide on action according to the inspection history space which expands in an exponential manner. Tactics suggest addressing this issue necessitates huge memory and can only perform properly for tiny, detached inspection spaces. Secondly, when presenting belief state to POMDP dilemmas, the belief space would not expand in an exponential way, but the knowledge of the model comes to be indispensable for the agent, which is not improper for numerous complex situations. Lastly, most solutions in POMDP quandaries should encounter a challenge known as the information gathering and exploitation dilemma. In a POMDP, the agent does not realize the present state precisely. It should determine whether to acquire extra information regarding the real state first or to utilize its present knowledge first. Definitely, for discovering the optimum policy, an agent should have extra environmental interactions.

Aside from the abovementioned POMDP challenges, the DRL model formulation and parameter optimization for various IoT applications are different from one case to another. Furthermore, a more effective solution might be developed based on the particular attributes of the IoT environment [74].

Joint reward from multiple agents: It was discussed that some efforts considered numerous DRL agents situated at various sensors or IoT devices in a disseminated way. Whereas every agent is responsible for performing a certain job or a group of similar jobs. In general, many agents are believed to carry out similar tasks. The development of multi-agent DRL-based security solutions, in which various agents performing diverse tasks can be viewed as a promising direction to be investigated. The challenge might be the cooperative technique of contemplating the reward values from overall agents, which make up the design of multi-agents security solution a complicated and challenging issue. Furthermore, the dynamic nature of the IoT environment making the appropriate control between various agents is a challenging task to be investigated [73].

10.7   Privacy-Preserving Federated Learning

This section mainly introduces some suggestions and directions to investigate as promising future work for more efficient privacy-preserving deep learning approaches:

A fascinating opening viewpoint, in the direction of future development of deep learning-based security solutions, would be by taking benefits from the cutting-edge research and discovering the diverse learned lessons drawn from the previous chapters. This includes investigating possible hybridization between various paradigms and their underlying technologies. A performance study in terms of efficiency, efficacy, and privacy should be conducted for theoretically auspicious hybridization. Then again, it is noteworthy that developing a real-world solution necessitates tackling the encountered privacy concerns as a whole problem from their various facets, this might necessitate a mixture of various privacy preservation solutions. For instance, developing a security solution enabling cooperative training of public federated deep learning at cloud, could necessitate not only protecting the participants and their training data throughout the learning phase but also preventing the resultant model from leaking private information to other participants or clients. Therefore, investigating such a solution is promising to the direction to protect privacy during the learning and release stage of the deep learning model.

Additionally, though a variety of attempts were presented to address the privacy protection of deep learning either for model training, usage or distributing, however generalizing one solution or absolutely favoring any solution upon others could not be viewed as the most proper notion. The development of a positive solution that realizes together suitable efficacy, efficiency, and privacy be contingent on diverse confines and necessities such as communication charges, time restraints, or accessible resources for representative users' devices to accomplish encryption/decryption for example, or to implement and run a local feature extractor when user accords [9]. Nevertheless, it is fascinating to develop adaptable solutions (federated or centralized) that permit a particular trade-off including efficiency, effectiveness, secrecy, and capacity to tackle constrictions and needs of various IoT applications.

Moreover, pledging future prospects in the privacy-preserving deep learning approach, especially in the perspective of cooperative or federated model learning, could also entail additional discovery of the potential of Blockchain as an evolving technology these days. Blockchain, originally utilized in the monetary industry, is fundamentally defined as a peer-to-peer distributed ledger technology or database embodied by a succession of data blocks secured and connected making use of cryptography [66–68]. Some research efforts recently introduced to develop a blockchain-related solution for tackling privacy preservation in deep learning [66–70]. Some research efforts introduced a secure distributed deep learning framework that employ a value-guided incentive strategy on the basis of Blockchain to tackle the problem of malevolent adversaries along with the absence of inducements for mistrustful participants. These efforts pave the way toward more investigation and development of blockchain-protected deep learning solutions in both the academic ad industrial domain.

From a hardware standpoint, it worth noting that despite the usage of trustworthy computer hardware could render a privacy-preserving machine or deep learning more feasible for real-world IoT applications, but complete confidentiality remains unguaranteed. However, it was indicated that this matter was fundamentally caused by the implementation shortcomings paving the way toward more discovery of this direction. In Addition, innovative processing technology dedicated large efforts for accelerating the deep learning models using Neural Processing Units such as Tensor Processor Unit (TPU) from Google, Neural network processors (**NNP**), and Myriad Development Kit (MDK) from Intel, Amazon Web Services Inferential from Amazon, or Nvidia Deep Learning Accelerator (NVDLA) from Nvidia, etc. This could be beneficial on the way to improved support of privacy techniques necessitating local deep learning processing or even the development of more effective techniques. Moreover, studying in what way to utilize Neural Processing Units for non-neural-network processing or hybrid models (e.g., fuzzy deep learning, evolutionary deep learning, etc.) can be viewed as an interesting direction to investigate and contrasted with other robust and prevailing processors, such as Graphical processing units (GPU), from the perspective of privacy-preserved deep learning solutions through IoT networks.

Finally, the problems of computational burdens and communication costs at the edge of IoT networks are likely to be tackled by investigating compression techniques for deep learning models, which can be viewed as an exciting future prospect to pursue according to the recent stated auspicious results [61]. Some research effort [75] demonstrated that a three-stage approach (containing pruning, trained quantization, and Huffman coding), is capable of reducing the storage prerequisite of a deep learning model by $35 \times$ to $49 \times$ without influencing their precision.

- Sparsification-Empowered Federated Learning

What is the optimal for developing federated learning solutions in a highly heterogeneous environment with severely restrained wireless resources? sparsification can
be viewed as an ideal way of enabling the design of a federated learning framework for a large number of heterogeneous participants (IoT devices). The sparsification enables a subset of participants to upload local parameters of a locally trained
deep learning model to a centralized authority server. The principle for selecting
the set of participants to upload their local parameters to the parameter server is a
thought-provoking mission. The simplest way for selecting the participant is to apply
random sampling of overall participants, however, this randomized solution might
not guarantee reaching optimal performance. In another way, the subset of participants can be selected according to the condition whether the value of local gradients
exceeds a predefined threshold or not. Nevertheless, it is difficult to calculate the
ideal threshold constant for a huge number of dissimilar participants. An Additional
principle for the assortment of participants can be the performance of their local deep
learning model, which is determined based on several features such as the quality of
data, computing resources, size of data, backup power, etc. Furthermore, the authors
with low performance over wireless channels cause extending the convergence time
of federated learning because of elevated packet error rates [76, 77]. To discourse
these challenges, more efforts should be dedicated to more intelligent and operative
sparsification-empowered federated learning protocols.

Though sparsification-Empowerd federated learning solution could lessen the
utilization of communication resources, however, it may not be adequate alone for
lots of participants' devices and constrained communication resources. To tackle this
challenge, quantization-enabled federated learning can be viewed as an encouraging
approach. Nevertheless, quantization engenders some inaccuracies in the global
federated framework because of the signal falsification. This kind of inaccuracy
in the trained federated model may possibly extend its convergence period. Consequently, it is highly suggested to develop a new federated learning scheme to take
into consideration the trade-off between utilization of communication resources and
convergence rate throughout the training phase. Numerous quantization methods
might be applied for federated learning across IoT networks including hyper-sphere
quantization [78], low precision quantizer [79, 80], and universal vector quantization
[81].

- Heterogeneous-Clustering Federated Learning

In what way the federated learning could be enabled for countless participants experiencing heterogeneous features? a large number of federated learning participants
often demonstrate a considerable volume of statistical heterogeneity in the corresponding local data. FedAvg algorithm was developed to calculate the parameters
of the global federated learning model based on the hypothesis of the synchronous
quantity of labor at participants' side during each communication round [19]. Nevertheless, the participants in IoT environments often showing considerable heterogeneity in their local data and learning parameters. In an attempt to tackle this

problem, FedProx was introduced to include a subjective proximal term into the federated model of FedAvg [76]. The process of deciding the weights for FedProx based on the intended application scenario is regarded difficult task. In Addition, nothing can guarantee that FedProx could incontrovertibly enhance the convergence ratio. Thus, it is highly recommended to develop an innovative federated learning protocol to handle IoT data heterogeneity issues in the context of privacy-preserved learning. One potential way to address that is the establishment of clusters of participants based on statistical uniformity. At each cluster, a cluster leader is chosen to take on the responsibility of aggregating the learning parameters submitted by local participants. Whilst the choice of cluster leaders ought to pursue some logical and challenging principles. For example, the principle for group leader assortment might improve in the global cluster's output. An Additional conceivable way could be publicly informed clustering. As cluster leader is able to deduce the participants' confidential information from the gradient of the local deep learning model [77], it may be advantageous to create cluster leader with superior social confidence nature. Within each cluster, a sub-universal model could be calculated in comparable way as with conventional federated learning. Thereafter, cluster leaders upload the parameters of their sub-universal deep learning models to a universal parameter server, at which the parameters are aggregated to generate a final federated model, which is later distributed back to the cluster leaders. After All, the cluster leaders distribute the final model updates to the participants belonging to their clusters.

- Mobility-Informed Federated Learning

In What Way a smooth communication can be enabled between mobile participants with the central authority (i.e., fog or cloud server) throughout the federated training? In mobility-based IoT systems (i.e., smart transportations), the federated training involves a set of mobile participants (i.e., IoT devices, mobile edge nodes, a mix of them) that typically connected to the centralized server. Nevertheless, the mobility nature of federated participants could reason some defeat in coverage and therefore, result in degrading the performance of federated solutions. Numerous solutions might be introduced to address such downside. For instance, modifying the federated learning practices to take into account the mobility of participants as well as other related considerations throughout the participants' selection stage. The stationary participants or low mobility participants ought to be chosen or should have a higher priority during the selection. This solution looks to be encouraging in the circumstances where the number of participants necessary for accomplishing federated learning has partial mobility or even no mobility at all [82–84]. Nevertheless, in the case of highly mobile participants, it is essential to foresee the mobility degree of participants in the course of the selection process. Therefore, deep learning can be employed for mobility prediction as an essential future prospect.

- Homomorphic Encryption-Enabled Federated Learning

What is the secure way for exchanging the parameters between participants and aggregation server in federated learning scenarios?

While federated learning has been achieving great success in realizing privacy-preserved on-site distributed deep learning solutions, however, it has come up with a set of security issues and many privacy challenges. For instance, A malevolent participant might get access to the learning parameters throughout the communication between the aggregation server and participants for parameters exchange purposes. Subsequently, the malevolent participant might illegally modify the learning parameters or infer some information about the training data. Additionally, the gradient information may be exploited by a malevolent participant to deduce confidential information about other participants [85–89]. Thus, it is essential to guarantee secure parameters exchange over communication channels. As a remedy, homomorphic encryption can be exploited for guaranteeing the secure sharing of parameters during federated training, where a key-pair is orchestrated between all participants over a secure channel. Specifically, every participant encrypts the local training parameters to be uploaded as ciphertexts to a centralized server. Then, the parameter server broadcasts the updated global parameters to all participants. Some homomorphic encryptions necessitate decryption at the aggregation server some other do not. However, commonly, algorithms of homomorphic encryption are complex necessitate the need for excessive computational and communicational IoT resources. The homomorphic encryption algorithms might be partitioned into three core categories involving partial algorithms, somewhat algorithms, and full algorithms encryption. These types of homomorphic encryption present the liberty of applying arithmetic operations dependent on the category of the algorithm is applied. For instance, partial encryption enables one and only one form of arithmetic calculations (i.e., multiplication or summation), whereas a full encryption algorithm permits countless distinct forms of arithmetic calculations. Moreover, the homomorphic encryption-generated ciphertext includes some level of noise that rises proportionately with arithmetic procedures [87]. Consequently, it is highly recommended to develop a more improved and lightweight homophobic encryption algorithm that take into consideration such kind of noise.

- Secure and Trustworthy Federated Aggregation

In the context of federated learning, what is the secure way for aggregating the local parameters at the parameter server?

A malevolent participant might upload erroneous local parameters to the parameter server during the aggregation phase with the main aim to decelerate the convergence ratio of the federated learning model. Under certain conditions, the global model may possibly not converge because of the incidence of a malevolent participant. In view of this, it is highly recommended to validate the participants' local updates prior to contemplating them in the parameter aggregation process. Numerous solutions might be developed to empower protected parameters aggregation. Some efforts were dedicated to developing a secure aggregation technique [90], that achieved security for 1:3 of all participants but at the expense of additional complexity and communication overhead. However, the insecure set of participants might expose the aggregation process to a malicious participant, thus it is essential to investigate the design of a near 100% secure aggregation algorithm while keeping little communication or computation overhead. In addition, a reputation-based scheme was proposed to estimate the level of trustworthy contribution of participants in the federated training, whereas a consortium blockchain was applied to support effective managing of reputation. Despite the success of blockchain-based solutions in protecting the aggregation against adversaries, however, they, unfortunately, exhibit high-latency problems limiting their real-world applicability [91–93]. To tackle these challenges, it is believed that investigating a new lightweight scheme for federated secure and trustworthy aggregation is a promising future prospect.

- Resource Intervention in Federated Learning

In the context of federated learning, what the secure way to reuse the uplink communication resources already under usage of other cellular operators? To secure the cellular participants from intervention during involvement in federated learning, resource allocation policies will be required. Two distinct strategies of distributing resource units to the federated participants. The first one is to allocate a specific resource unit to just a single participant while satisfying the restraint of the highest permitted intervention. Another strategy seeks to allocate manifold participants to a one resource unit to enable efficient and effective resource recycling. In the former circumstance, a one-to-one matching scheme is employed, while in the latter circumstance, a one-to-many matching scheme with outwardness for resource allocation because of the datum that favorite profiles rely on resource units as well as other participants allocated to a single resource unit [94]. Furthermore, for a static allocation of resource units and static transmit power, the participants' interrelation has no effect on intrusion management for cellular participants whilst recycling the corresponding frequencies. Nevertheless, mutual allocation of transmit power and device interconnectivity of participants' devices could be accomplished to concurrently lessen the charge of federated learning and cellular intervention. This type of cooperative broadcast power distribution and the device-association optimization challenge can be formulated as a mixed-integer non-linear programming problem, which can be resolved in different ways.

- Adaptive Heterogeneous Resource Allocation

In the context of federated Learning, how to train a federated model with a small convergence rate for participants having a heterogeneous set of resources? The federated learning participants are often accompanied by a heterogeneous set of resources i.e., computing, storage, power, and communication resources. This heterogeneity typically influences federated learning performance. For example, the time complexity of local deep learning models rigorously varies with local performance, local operational frequency, and the size of local training data. Besides, the synchronous nature of federated learning stems from completing the local model training within static time intervals. Nevertheless, the heterogeneity of participants' devices makes the local training time witness considerable variations resulting in different convergence rates and thereby different local performances. Consequently, the issues related to resource heterogeneity have to be considered in future works. As a possible solution, the impact of heterogeneity can be diminished by allocating extra computational resources to participants having bigger amounts of local training data to realize acceptable performance within the permitted time interval and contrariwise [95, 96]. In another way, a method to allocate extra communication resources to the participants with weak performance because of shortage in computing resources. A promising future prospect might include adaptive allocation of resources participants based on underlying characters with the main objective to fasten the convergence of federated models while maintaining robust performance.

References

1. I. Makhdoom, M. Abolhasan, J. Lipman, R.P. Liu, W. Ni, Anatomy of threats to the internet of things. IEEE Commun. Surv. Tutorials (2019). https://doi.org/10.1109/COMST.2018.287 4978
2. N. Asokan, F. Brasser, A. Ibrahim, A.R. Sadeghi, M. Schunter, G. Tsudik, C. Wachsmann, SEDA: scalable embedded device attestation, in: Proc. ACM Conf. Comput. Commun. Secur. (2015). https://doi.org/10.1145/2810103.2813670
3. S. Mubeen, S.A. Asadollah, A.V. Papadopoulos, M. Ashjaei, H. Pei-Breivold, M. Behnam, Management of service level agreements for cloud services in IoT: a systematic mapping study. IEEE Access (2018). https://doi.org/10.1109/ACCESS.2017.2744677
4. N. Agrawal, S. Tapaswi, Defense mechanisms against DDoS attacks in a cloud computing environment: state-of-the-art and research challenges, IEEE Commun. Surv. Tutorials (2019). https://doi.org/10.1109/COMST.2019.2934468
5. S. Dong, K. Abbas, R. Jain, A survey on distributed denial of service (DDoS) attacks in SDN and cloud computing environments. IEEE Access (2019). https://doi.org/10.1109/ACCESS. 2019.2922196
6. W. Zhijun, L. Wenjing, L. Liang, Y. Meng, Low-rate DoS attacks, detection, defense, and challenges: a survey. IEEE Access (2020). https://doi.org/10.1109/ACCESS.2020.2976609
7. N. Srihari Rao, K. Chandra Sekharaiah, A. Ananda Rao, A survey of distributed denial-of-service (DDoS) Defense techniques in ISP domains, in: Lect. Notes Networks Syst. (2019). https://doi.org/10.1007/978-981-10-8201-6_25
8. X. Jing, Z. Yan, X. Jiang, W. Pedrycz, Network traffic fusion and analysis against DDoS flooding attacks with a novel reversible sketch, Inf. Fusion. (2019). https://doi.org/10.1016/j. inffus.2018.10.013
9. M.A. Sotelo Monge, A. Herranz González, B. Lorenzo Fernández, D. Maestre Vidal, G. Rius García, J. Maestre Vidal, Traffic-flow analysis for source-side DDoS recognition on 5G environments. J. Netw. Comput. Appl. (2019). https://doi.org/10.1016/j.jnca.2019.02.030
10. S.Q. Ali Shah, F. Zeeshan Khan, M. Ahmad, The impact and mitigation of ICMP based economic denial of sustainability attack in cloud computing environment using software defined network. Comput. Networks. (2021). https://doi.org/10.1016/j.comnet.2021.107825
11. P.T. Dinh, M. Park, R-EDoS: robust economic denial of sustainability detection in an sdn-based cloud through stochastic recurrent neural network. IEEE Access (2021). https://doi.org/ 10.1109/ACCESS.2021.3061601
12. Q. Yan, F.R. Yu, Q. Gong, J. Li, Software-defined networking (SDN) and distributed denial of service (DDOS) attacks in cloud computing environments: a survey, some research issues, and challenges. IEEE Commun. Surv. Tutorials. (2016). https://doi.org/10.1109/COMST.2015.248 7361
13. G. Somani, M.S. Gaur, D. Sanghi, M. Conti, M. Rajarajan, R. Buyya, Combating DDoS attacks in the cloud: requirements, trends, and future directions. IEEE Cloud Comput. (2017). https:// doi.org/10.1109/MCC.2017.14

14. J.J. Santanna, J. de Vries, R. de O. Schmidt, D. Tuncer, L. Z. Granville, A. Pras, Booter list generation: the basis for investigating DDoS-for-hire websites, in: Int. J. Netw. Manag. (2018). https://doi.org/10.1002/nem.2008

15. M. Stoyanova, Y. Nikoloudakis, S. Panagiotakis, E. Pallis, E.K. Markakis, A survey on the internet of things (IoT) forensics: challenges, approaches, and open issues. IEEE Commun. Surv. Tutorials. (2020). https://doi.org/10.1109/COMST.2019.2962586

16. M. Conti, A. Dehghantanha, K. Franke, S. Watson, Internet of Things security and forensics: challenges and opportunities. Futur. Gener. Comput. Syst. (2018). https://doi.org/10.1016/j.fut ure.2017.07.060

17. P. Hu, S. Dhelim, H. Ning, T. Qiu, Survey on fog computing: architecture, key technologies, applications and open issues. J. Netw. Comput. Appl. (2017). https://doi.org/10.1016/j.jnca. 2017.09.002

18. K. Tange, M. De Donno, X. Fafoutis, N. Dragoni, A systematic survey of industrial internet of things security: requirements and fog computing opportunities. IEEE Commun. Surv. Tutorials. (2020). https://doi.org/10.1109/COMST.2020.3011208

19. Y.I. Alzoubi, V.H. Osmanaj, A. Jaradat, A. Al-Ahmad, Fog computing security and privacy for the Internet of Thing applications: state-of-the-art, Secur. Priv. (2021). https://doi.org/10. 1002/spy2.145

20. M. Mukherjee, L. Shu, D. Wang, Survey of fog computing: fundamental, network applications, and research challenges. IEEE Commun. Surv. Tutorials. (2018). https://doi.org/10.1109/ COMST.2018.2814571

21. P.Y. Zhang, M.C. Zhou, G. Fortino, Security and trust issues in Fog computing: A survey, Futur. Gener. Comput. Syst. (2018). https://doi.org/10.1016/j.future.2018.05.008

22. P.K. Das, D. Ghosh, P. Jagtap, A. Joshi, T. Finin, Preserving user privacy and security in context-aware mobile platforms, in: Censorship, Surveillance, Priv. Concepts, Methodol. Tools, Appl. (2018). https://doi.org/10.4018/978-1-5225-7113-1.ch058

23. H. Lin, Z. Yan, Y. Fu, Adaptive security-related data collection with context awareness. J. Netw. Comput. Appl. (2019). https://doi.org/10.1016/j.jnca.2018.11.002

24. P. Ranaweera, V.N. Imrith, M. Liyanage, A.D. Jurcut, Security as a service platform leveraging multi-access edge computing infrastructure provisions. IEEE Int. Conf. Commun., (2020). https://doi.org/10.1109/ICC40277.2020.9148660

25. P. Ranaweera, A.D. Jurcut, M. Liyanage, Survey on multi-access edge computing security and privacy. IEEE Commun. Surv. Tutorials. (2021). https://doi.org/10.1109/COMST.2021. 3062546

26. A. Sill, The design and architecture of microservices. IEEE Cloud Comput. (2016). https://doi. org/10.1109/MCC.2016.111

27. Z. Wen, T. Lin, R. Yang, S. Ji, R. Ranjan, A. Romanovsky, C. Lin, J. Xu, GA-Par: dependable microservice orchestration framework for geo-distributed clouds. IEEE Trans. Parallel Distrib. Syst. (2020). https://doi.org/10.1109/TPDS.2019.2929389

28. M. Villari, M. Fazio, S. Dustdar, O. Rana, R. Ranjan, Osmotic computing: a new paradigm for edge/cloud integration. IEEE Cloud Comput. (2016). https://doi.org/10.1109/MCC.2016.124

29. A. Buzachis, A. Galletta, A. Celesti, L. Carnevale, M. Villari, Towards osmotic computing: a blue-green strategy for the fast re-deployment of microservices. Proc.—IEEE Symp. Comput. Commun., (2019). https://doi.org/10.1109/ISCC47284.2019.8969621

30. T. Taleb, K. Samdanis, B. Mada, H. Flinck, S. Dutta, D. Sabella, On Multi-access edge computing: a survey of the emerging 5g network edge cloud architecture and orchestration. IEEE Commun. Surv. Tutorials. (2017). https://doi.org/10.1109/COMST.2017.2705720

31. W. Rafique, L. Qi, I. Yaqoob, M. Imran, R.U. Rasool, W. Dou, Complementing IoT services through software defined networking and edge computing: a comprehensive survey. IEEE Commun. Surv. Tutorials. (2020). https://doi.org/10.1109/COMST.2020.2997475

32. Y. Wu, Cloud-edge orchestration for the internet-of-things: architecture and AI-powered data processing. IEEE Internet Things J. (2020). https://doi.org/10.1109/JIOT.2020.3014845

33. Z.M. Fadlullah, F. Tang, B. Mao, N. Kato, O. Akashi, T. Inoue, K. Mizutani, State-of-the-Art deep learning: evolving machine intelligence toward tomorrow's intelligent network traffic control systems. IEEE Commun. Surv. Tutorials. (2017). https://doi.org/10.1109/COMST. 2017.2707140

34. Q. Zhang, L.T. Yang, Z. Chen, P. Li, A survey on deep learning for big data. Inf. Fusion. (2018). https://doi.org/10.1016/j.inffus.2017.10.006
35. M.M. Najafabadi, F. Villanustre, T.M. Khoshgoftaar, N. Seliya, R. Wald, E. Muharemagic, Deep learning applications and challenges in big data analytics. J. Big Data. (2015). https://doi.org/10.1186/s40537-014-0007-7
36. M. Mohammadi, A. Al-Fuqaha, S. Sorour, M. Guizani, Deep learning for IoT big data and streaming analytics: a survey, IEEE Commun. Surv. Tutorials. (2018). https://doi.org/10.1109/COMST.2018.2844341
37. M. Ge, H. Bangui, B. Buhnova, Big data for Internet of Things: a survey. Futur. Gener. Comput. Syst. (2018). https://doi.org/10.1016/j.future.2018.04.053
38. M. Pesteie, P. Abolmaesumi, R.N. Rohling, Adaptive augmentation of medical data using independently conditional variational auto-encoders. IEEE Trans. Med. Imaging. (2019). https://doi.org/10.1109/TMI.2019.2914656
39. H.F. Nweke, Y.W. Teh, M.A. Al-garadi, U.R. Alo, Deep learning algorithms for human activity recognition using mobile and wearable sensor networks: State of the art and research challenges. Expert Syst. Appl. (2018). https://doi.org/10.1016/j.eswa.2018.03.056
40. T.T. Um, F.M.J. Pfister, D. Pichler, S. Endo, M. Lang, S. Hirche, U. Fietzek, D. Kulic, Data augmentation of wearable sensor data for Parkinson's disease monitoring using convolutional neural networks. ICMI 2017—Proc. 19th ACM Int. Conf. Multimodal Interact., (2017). https://doi.org/10.1145/3136755.3136817
41. P.M. Comar, L. Liu, S. Saha, P.N. Tan, A. Nucci, Combining supervised and unsupervised learning for zero-day malware detection. Proc.—IEEE INFOCOM, (2013). https://doi.org/10.1109/INFCOM.2013.6567003
42. H. Hindy, R. Atkinson, C. Tachtatzis, J.N. Colin, E. Bayne, X. Bellekens, Utilising deep learning techniques for effective zero-day attack detection. Electron. (2020). https://doi.org/10.3390/electronics9101684
43. M. Zhang, L. Wang, S. Jajodia, A. Singhal, Network attack surface: lifting the concept of attack surface to the network level for evaluating networks' resilience against zero-day attacks, IEEE Trans. Dependable Secur. Comput. (2021). https://doi.org/10.1109/TDSC.2018.2889086
44. N. Chaabouni, M. Mosbah, A. Zemmari, C. Sauvignac, P. Faruki, Network intrusion detection for IoT security based on learning techniques. IEEE Commun. Surv. Tutorials. (2019). https://doi.org/10.1109/COMST.2019.2896380
45. X. Zhang, O. Upton, N.L. Beebe, K.K.R. Choo, IoT Botnet Forensics: A comprehensive digital forensic case study on mirai botnet servers. Forensic Sci. Int. Digit. Investig. (2020). https://doi.org/10.1016/j.fsidi.2020.300926
46. C. Frank, C. Nance, S. Jarocki, W.E. Pauli, S.D. Madison, Protecting IoT from Mirai botnets; IoT device hardening. Proc. Conf. Inf. Syst. Appl. Res. ISSN. (2017)
47. H. Griffioen, C. Doerr, Examining Mirai's Battle over the Internet of Things. Proc. ACM Conf. Comput. Commun. Secur., (2020). https://doi.org/10.1145/3372297.3417277
48. F. Ali Garba, S. B. Junaidu, A. A. Obiniyi, A. M. Ibrahim, Improved Mirai Bot scanner summation algorithm. SSRN Electron. J. (2020). https://doi.org/10.2139/ssrn.3519728
49. R.S. Abdullah, A. Nazri, W. Yasin, M.A. Faizal, W.N.F.W. Mohd Zaki, Constructing pattern of mirai botnets attack using graph theory approach. Int. J. Adv. Trends Comput. Sci. Eng. (2020). https://doi.org/10.30534/ijatcse/2020/55952020
50. G.I. Parisi, R. Kemker, J.L. Part, C. Kanan, S. Wermter, Continual lifelong learning with neural networks: a review. Neural Networks. (2019). https://doi.org/10.1016/j.neunet.2019.01.012
51. Y. Yue, S. Li, P. Legg, F. Li, Deep learning-based security behaviour analysis in IoT environments: a survey. Secur. Commun. Networks., (2021). https://doi.org/10.1155/2021/8873195
52. P. Li, Z. Chen, L.T. Yang, J. Gao, Q. Zhang, M.J. Deen, An incremental deep convolutional computation model for feature learning on industrial big data. IEEE Trans. Ind. Informatics., (2019). https://doi.org/10.1109/TII.2018.2871084
53. Q. Zhang, L.T. Yang, Z. Chen, P. Li, Incremental deep computation model for wireless big data feature learning. IEEE Trans. Big Data., (2019). https://doi.org/10.1109/tbdata.2019.2903092

54. S. Huang, Y. Guo, D. Liu, S. Zha, W. Fang, A two-stage transfer learning-based deep learning approach for production progress prediction in IoT-enabled manufacturing. IEEE Internet Things J. (2019). https://doi.org/10.1109/JIOT.2019.2940131

55. L. Vu, Q.U. Nguyen, D.N. Nguyen, D.T. Hoang, E. Dutkiewicz, Deep Transfer Learning for IoT Attack Detection, IEEE Access. (2020). https://doi.org/10.1109/ACCESS.2020.3000476

56. S.R. Pokhrel, L. Pan, N. Kumar, R. Doss, H.L. Vu, Multipath TCP meets transfer learning: a novel edge-based learning for industrial IoT. IEEE Internet Things J. (2021). https://doi.org/10.1109/jiot.2021.3056466

57. C.H. Lu, X.Z. Lin, Toward direct edge-to-edge transfer learning for IoT-enabled edge cameras. IEEE Internet Things J., (2021). https://doi.org/10.1109/JIOT.2020.3034153

58. M.A. Al-Garadi, A. Mohamed, A.K. Al-Ali, X. Du, I. Ali, M. Guizani, A survey of machine and deep learning methods for Internet of Things (IoT) security. IEEE Commun. Surv. Tutorials., (2020). https://doi.org/10.1109/COMST.2020.2988293

59. S. Srivastava, A. Bhatia, On the learning capabilities of recurrent neural networks: a cryptographic perspective. Proc.—9th IEEE Int. Conf. Big Knowledge, ICBK 2018, (2018). https://doi.org/10.1109/ICBK.2018.00029

60. B. Hitaj, G. Ateniese, F. Perez-Cruz, Deep models under the GAN: information leakage from collaborative deep learning. Proc. ACM Conf. Comput. Commun. Secur., (2017). https://doi.org/10.1145/3133956.3134012

61. A. Boulemtafes, A. Derhab, Y. Challal, A review of privacy-preserving techniques for deep learning. Neurocomputing. (2020). https://doi.org/10.1016/j.neucom.2019.11.041

62. J. Zhang, B. Chen, X. Cheng, H.T.T. Binh, S. Yu, PoisonGAN: generative poisoning attacks against federated learning in edge computing systems. IEEE Internet Things J. (2021). https://doi.org/10.1109/JIOT.2020.3023126

63. W. Jiang, H. Li, S. Liu, X. Luo, R. Lu, Poisoning and evasion attacks against deep learning algorithms in autonomous vehicles. IEEE Trans. Veh. Technol., (2020). https://doi.org/10.1109/TVT.2020.2977378

64. S. Tu, M. Waqas, S.U. Rehman, T. Mir, G. Abbas, Z.H. Abbas, Z. Halim, I. Ahmad, Reinforcement learning assisted impersonation attack detection in device-to-device communications. IEEE Trans. Veh. Technol. (2021). https://doi.org/10.1109/TVT.2021.3053015

65. N. Subbanna, M. Wilms, A. Tuladhar, N.D. Forkert, An analysis of the vulnerability of two common deep learning-based medical image segmentation techniques to model inversion attacks. Sensors. (2021). https://doi.org/10.3390/s21113874

66. Y. Alufaisan, M. Kantarcioglu, Y. Zhou, Robust transparency against model inversion attacks. IEEE Trans. Dependable Secur. Comput., (2020). https://doi.org/10.1109/TDSC.2020.3019508

67. S. Rathore, J.H. Park, A blockchain-based deep learning approach for cyber security in next generation industrial cyber-physical systems. IEEE Trans. Ind. Informatics. (2021). https://doi.org/10.1109/TII.2020.3040968

68. M. Singh, G.S. Aujla, R.S. Bali, A deep learning-based blockchain mechanism for secure internet of drones environment. IEEE Trans. Intell. Transp. Syst., (2020). https://doi.org/10.1109/tits.2020.2997469

69. M. Singh, G.S. Aujla, A. Singh, N. Kumar, S. Garg, Deep-learning-based blockchain framework for secure software-defined industrial networks. IEEE Trans. Ind. Informatics., (2021). https://doi.org/10.1109/TII.2020.2968946

70. O. Valery, P. Liu, J.J. Wu, A collaborative CPU-GPU approach for deep learning on mobile devices. Concurr. Comput., (2019). https://doi.org/10.1002/cpe.5225

71. S. Wang, G. Ananthanarayanan, T. Mitra, OPTiC: optimizing collaborative CPU-GPU computing on mobile devices with thermal constraints. IEEE Trans. Comput. Des. Integr. Circuits Syst., (2019). https://doi.org/10.1109/TCAD.2018.2873210

72. S. Alharby, N. Harris, A. Weddell, J. Reeve, The security trade-offs in resource constrained nodes for IoT application. Int. J. Electr. Electron. Commun. Sci., **11**.0. (2018)

73. A. Uprety, D.B. Rawat, Reinforcement learning for IoT Security: A Comprehensive Survey. IEEE Internet Things J., (2021). https://doi.org/10.1109/JIOT.2020.3040957

74. L. Lei, Y. Tan, K. Zheng, S. Liu, K. Zhang, X. Shen, Deep reinforcement learning for autonomous internet of things: model, applications and challenges. IEEE Commun. Surv. Tutorials., (2020). https://doi.org/10.1109/COMST.2020.2988367
75. S. Han, H. Mao, W.J. Dally, Deep compression: Compressing deep neural networks with pruning, trained quantization and Huffman coding. 4th Int. Conf. Learn. Represent. ICLR 2016—Conf. Track Proc., (2016)
76. H. Sun, S. Li, F. Richard Yu, Q. Qi, J. Wang, J. Liao, Toward communication-efficient federated learning in the internet of things with edge computing. IEEE Internet Things J., (2020). https://doi.org/10.1109/JIOT.2020.2994596
77. L.U. Khan, W. Saad, Z. Han, E. Hossain, C.S. Hong, Federated learning for internet of things: recent advances, taxonomy, and open challenges. IEEE Commun. Surv. Tutorials. (2021). https://doi.org/10.1109/comst.2021.3090430
78. S. Eghbali, L. Tahvildari, Deep spherical quantization for image search, in: Proc. IEEE Comput. Soc. Conf. Comput. Vis. Pattern Recognit., (2019). https://doi.org/10.1109/CVPR.2019.01196
79. S. Jung, C. Son, S. Lee, J. Son, Y. Kwak, J.-J. Han, C. Choi, Joint training of low-precision neural network with quantization interval parameters. Rev. Int. La Croix-Rouge Bull. Int. Des Sociétés La Croix-Rouge., (2018)
80. M. Rusci, M. Fariselli, A. Capotondi, L. Benini, Leveraging automated mixed-low-precision quantization for tiny edge microcontrollers. Commun. Comput. Inf. Sci., (2020). https://doi.org/10.1007/978-3-030-66770-2_22
81. N. Shlezinger, M. Chen, Y.C. Eldar, H.V. Poor, S. Cui, UVeQFed: universal vector quantization for federated learning. IEEE Trans. Signal Process., (2021). https://doi.org/10.1109/TSP.2020.3046971
82. Z. Yu, J. Hu, G. Min, Z. Zhao, W. Miao, M.S. Hossain, Mobility-aware proactive edge caching for connected vehicles using federated learning. IEEE Trans. Intell. Transp. Syst., (2020). https://doi.org/10.1109/tits.2020.3017474
83. J. Kang, Z. Xiong, D. Niyato, Y. Zou, Y. Zhang, M. Guizani, Reliable federated learning for mobile networks. IEEE Wirel. Commun., (2020). https://doi.org/10.1109/MWC.001.1900119
84. W.Y.B. Lim, N.C. Luong, D.T. Hoang, Y. Jiao, Y.C. Liang, Q. Yang, D. Niyato, C. Miao, Federated Learning in Mobile Edge Networks: A Comprehensive Survey, IEEE Commun. Surv. Tutorials., (2020). https://doi.org/10.1109/COMST.2020.2986024
85. C. Zhou, A. Fu, S. Yu, W. Yang, H. Wang, Y. Zhang, Privacy-preserving federated learning in fog computing. IEEE Internet Things J., (2020). https://doi.org/10.1109/JIOT.2020.2987958
86. W. Ou, J. Zeng, Z. Guo, W. Yan, D. Liu, S. Fuentes, A homomorphic-encryption-based vertical federated learning scheme for rick management. Comput. Sci. Inf. Syst., (2020). https://doi.org/10.2298/CSIS190923022O
87. A. Acar, H. Aksu, A.S. Uluagac, M. Conti, A survey on homomorphic encryption schemes: theory and implementation. ACM Comput. Surv., (2018). https://doi.org/10.1145/3214303
88. A. Wood, K. Najarian, D. Kahrobaei, Homomorphic encryption for machine learning in medicine and bioinformatics. ACM Comput. Surv., (2020). https://doi.org/10.1145/3394658
89. F. Turan, S.S. Roy, I. Verbauwhede, HEAWS: an accelerator for homomorphic encryption on the amazon AWS FPGA. IEEE Trans. Comput., (2020). https://doi.org/10.1109/TC.2020.2988765
90. K. Bonawitz, V. Ivanov, B. Kreuter, A. Marcedone, H.B. McMahan, S. Patel, D. Ramage, A. Segal, K. Seth, Practical secure aggregation for privacy-preserving machine learning. Proc. ACM Conf. Comput. Commun. Secur., (2017). https://doi.org/10.1145/3133956.3133982
91. J. So, B. Guler, A.S. Avestimehr, Turbo-Aggregate: breaking the quadratic aggregation barrier in secure federated learning. IEEE J. Sel. Areas Inf. Theory., (2021). https://doi.org/10.1109/jsait.2021.3054610
92. D. Meng, H. Li, F. Zhu, X. Li, FedMONN: meta operation neural network for secure federated aggregation, in: Proc. 2020 IEEE 22nd Int. Conf. High Perform. Comput. Commun. IEEE 18th Int. Conf. Smart City IEEE 6th Int. Conf. Data Sci. Syst. HPCC-SmartCity-DSS 2020, (2020). https://doi.org/10.1109/HPCC-SmartCity-DSS50907.2020.00073

93. J. So, B. Guler, A.S. Avestimehr, Byzantine-resilient secure federated learning. IEEE J. Sel. Areas Commun., (2020). https://doi.org/10.1109/JSAC.2020.3041404

94. S.M.A. Kazmi, N.H. Tran, W. Saad, Z. Han, T.M. Ho, T.Z. Oo, C.S. Hong, Mode selection and resource allocation in device-to-device communications: a matching game approach. IEEE Trans. Mob. Comput., (2017). https://doi.org/10.1109/TMC.2017.2689768

95. L.U. Khan, S.R. Pandey, N.H. Tran, W. Saad, Z. Han, M.N.H. Nguyen, C.S. Hong, Federated learning for edge networks: resource optimization and incentive mechanism. IEEE Commun. Mag., (2020). https://doi.org/10.1109/MCOM.001.1900649

96. C. Shen, J. Xu, S. Zheng, X. Chen, Resource rationing for wireless federated learning: concept, benefits, and challenges. IEEE Commun. Mag., (2021). https://doi.org/10.1109/MCOM.001.2000744